T0320972

Digital Twin Technology

Digital Twin Technology

Edited by
Gopal Chaudhary, Manju Khari, and
Mohamed Elhoseny

CRC Press
Taylor & Francis Group
Boca Raton London New York

CRC Press is an imprint of the
Taylor & Francis Group, an **informa** business

First edition published 2022
by CRC Press
6000 Broken Sound Parkway NW, Suite 300, Boca Raton, FL 33487-2742

and by CRC Press
2 Park Square, Milton Park, Abingdon, Oxon, OX14 4RN

© 2022 Taylor & Francis Group, LLC

CRC Press is an imprint of Taylor & Francis Group, LLC

The right of Gopal Chaudhary, Manju Khari and Mohamed Elhoseny to be identified as the authors of the editorial material, and of the authors for their individual chapters, has been asserted in accordance with sections 77 and 78 of the Copyright, Designs and Patents Act 1988.

Reasonable efforts have been made to publish reliable data and information, but the author and publisher cannot assume responsibility for the validity of all materials or the consequences of their use. The authors and publishers have attempted to trace the copyright holders of all material reproduced in this publication and apologize to copyright holders if permission to publish in this form has not been obtained. If any copyright material has not been acknowledged please write and let us know so we may rectify in any future reprint.

Library of Congress Cataloging-in-Publication Data
A catalog record for this book has been requested

ISBN: 9780367677954 (hbk)
ISBN: 9780367677978 (pbk)
ISBN: 9781003132868 (ebk)

DOI: 10.1201/9781003132868

Typeset in Times
by codeMantra

Contents

Editors

Dr. Gopal Chaudhary is currently working as an assistant professor in Bharati Vidyapeeth's College of Engineering, Guru Gobind Singh Indraprastha University, Delhi, India. He holds a Ph.D. in biometrics at the division of Instrumentation and Control Engineering, Netaji Subhas Institute of Technology, University of Delhi, India. He received B.E. degree in electronics and communication engineering in 2009 and the MTech degree in microwave and optical communication from Delhi Technological University (formerly known as Delhi College of Engineering), New Delhi, India, in 2012. He has 30 publications in refereed national/international journals and conferences (e.g. Elsevier, Springer, Inderscience) in the area of biometrics and its applications. His current research interests include soft computing, intelligent systems, information fusion, and pattern recognition. He has organized many conferences and contributed to special issues.

Dr. Manju Khari is an assistant professor in Netaji Subhas University of Technology, East Campus, Delhi, India. She is also the professor-in-charge of the IT Services of the Institute and has experience of more than twelve years in network planning and management. She holds a Ph.D. in computer science & engineering from National Institute of Technology, Patna. She received her Master's degree in information security from Ambedkar Institute of Advanced Communication Technology and Research, formerly known as Ambedkar Institute of Technology affiliated to Guru Gobind Singh Indraprastha University, Delhi, India. Her research interests are software testing, information security, optimization, image processing, and machine learning. She has published more than 100 papers in refereed national/international journals and conferences (viz. IEEE, ACM, Springer, Inderscience, and Elsevier), and 6 book chapters in a Springer book. She is also a co-author of two books published by NCERT of Secondary and Senior Secondary School.

Dr. Mohamed Elhoseny is currently an assistant professor at the Faculty of Computers and Information, Mansoura University. Dr. Elhoseny has been appointed as an ACM Distinguished Speaker from 2019 to 2022. Collectively, Dr. Elhoseny authored/co-authored over 85 ISI journal articles in high-ranked and prestigious journals such as *IEEE Transactions on Industrial Informatics* (IEEE), *IEEE Transactions on Reliability* (IEEE), *Future Generation Computer Systems* (Elsevier), and *Neural Computing and Applications* (Springer). Besides, Dr. Elhoseny authored/ edited 15 international books (11 published by Springer, 2 published by Taylor& Francis, 1 published by Elsevier, and 1 published by IGI Global). His research interests include smart cities, network security, artificial intelligence, internet of things, and intelligent systems. Dr. Elhoseny serves as the Editor-in-Chief of *International Journal of Smart Sensor Technologies and Applications* (IGI Global). Moreover, he is an Associate Editor of many journals such as *IEEE Journal of Biomedical and Health Informatics* (IEEE), *IEEE Access* (IEEE), *Scientific Reports* (Nature),

IEEE Future Directions (IEEE), *Remote Sensing* (MDPI), and *International Journal of E-Services and Mobile Applications* (IGI Global), and *Applied Intelligence* (Springer). He also served as the co-chair, the publication chair, the program chair, and the track chair for several international conferences published by IEEE and Springer. Dr. Elhoseny is the Editor-in-Chief of the Studies in Distributed Intelligence Springer Book Series, the Editor-in-Chief of The Sensors Communication for Urban Intelligence CRC Press-Taylor& Francis Book Series, and the Editor-in-Chief of The Distributed Sensing and Intelligent Systems CRC Press-Taylor& Francis Book Series. He was granted several awards by diverse funding bodies such as the Young Researcher Award in Artificial Intelligence from the Federation of Arab Scientific Research Councils in 2019, Obada International Prize for young distinguished scientists 2020, the Egypt State Encouragement Award in 2018, the best Ph.D. thesis in Mansoura University in 2015, the SRGE best young researcher award in 2017, and the membership of The Egyptian Young Academy of Science (EYAS) in 2019. Besides, he is a TPC Member or Reviewer in 50+ international conferences and workshops. He has also been reviewing papers for 80+ international journals including *IEEE Communications Magazine*, *IEEE Transactions on Intelligent Transportation Systems*, *IEEE Sensors Letters*, *IEEE Communication Letters*, *Elsevier Computer Communications*, *Computer Networks*, *Sustainable Cities and Society*, *Wireless Personal Communications*, and *Expert Systems with Applications*. Dr. Elhoseny has been invited as a guest in many media programs to comment on technologies and related issues.

Contributors

Hadi Arabi
Department of Architecture
    Engineering
Pars University of Architecture and Art
Tehran, Iran

J. Arora
IT Department
Maharaja Surajmal Institute of
    Technology
New Delhi, India

Somayeh Asadi
Department of Architectural
    Engineering
Pennsylvania State University
State College, Pennsylvania

Sapna Dewari
Computer Science and Engineering
    Department
Chandigarh University
Punjab, India

Anjali Donkal
MCA Department
National Institute of Technology
Kurukshetra, India

Gita Donkal
CSE Department
Chandigarh University
Punjab, India

Meenu Gupta
Computer Science and Engineering
    Department
Chandigarh University
Punjab, India

Himanshu
Department of Computer Science and
    Engineering
Ambedkar Institute of Advanced
    Communication Technologies and
    Research
New Delhi, India

Vanita Jain
Department of Information Technology
Bharati Vidyapeeth's College of
    Engineering
New Delhi, India

Vibhuti Jain
Department of Computer Science
Guru Gobind Singh Indraprastha
    University
New Delhi, India

Jagannath Jayanti
Department of Computer Science
Guru Gobind Singh Indraprastha
    University
New Delhi, India

Manju Khari
School of Computer and System
    Sciences
Jawaharlal Nehru University
New Delhi, India

Mridul Khurana
Department of Information
    Technology
Bharati Vidyapeeth's College of
    Engineering
New Delhi, India

Charles J. Kibert
Director of Powell Center for
    Construction and Environment
University of Florida
Gainesville, Florida

Kapil Kumar
Computer Science and Engineering
Netaji Subhas University of Technology
    East Campus
New Delhi, India

Rakesh Kumar
Computer Science and Engineering
    Department
Chandigarh University
Mohali, India

M. Kumari
Department of ECE
Chandigarh University
Mohali, India

N. Luthra
Department of Information Technology
Bharati Vidyapeeth's College of
    Engineering
New Delhi, India

Fatemeh Mahdavi
Department of Conservation and
    Restoration
Tehran University of Art
Tehran, Iran

Hamed Niroumand
Department of Civil Engineering
Buein Zahra Technical University
    Education Center of Engineering and
    Technology
Imam Khomeini International
    University (IKIU)
Qazvin, Iran
and
Powell Center for Construction and
    Environment
University of Florida
Gainesville, Florida

Akanshu Raj
Department of Information Technology
Bharati Vidyapeeth's College of
    Engineering
New Delhi, India

D. Saini
Department of Computer Science
Bharati Vidyapeeth's College of
    Engineering
New Delhi, India

Divya Singh
Department of Computer Science
Computer Science and Engineering
    Department
Chandigarh University
Mohali, India

Madhavendra Singh
Department of Computer Science
Guru Gobind Singh Indraprastha
    University
New Delhi, India

Abhishek Tanwar
Department of Information Technology
Bharati Vidyapeeth's College of
 Engineering
New Delhi, India

M. Tushir
Department of Electrical & Electronics
 Engineering
Maharaja Surajmal Institute of
 Technology
New Delhi, India

C. Ved
Department of Information Technology
Bharati Vidyapeeth's College of
 Engineering
New Delhi, India

1 Digital Twin Technology
An Evaluation

Vanita Jain, N. Luthra, and D. Saini
Bharati Vidyapeeth's College of Engineering

CONTENTS

DOI: 10.1201/9781003132868-1

1.1 INTRODUCTION: BACKGROUND

As technology moves forward at lightning speed, it is really easy to overlook the complete impact of technology in today's world. As we have seen in automobiles' invention, it made it possible to travel long distances in hours rather than days. All inventions from the wheel to the compass, clock, steam engine, and several other inventions have made our lives simpler than even kings used to have in the past. It is believed that the invention of wheel made today's automobile possible. But if we look a couple of centuries back, the automobile was a fantasy. Similarly, the Internet is believed to be a new wheel that brings a radical change in our society. Already technological advancements have changed everything from how we consume information to economic transactions [1–3] taking place throughout the world. Millions of transactions are taking place every second, and a zettabyte of data is getting consumed [4,5]. As Steve Jobs said, "You can't connect the dots looking forward, you can only connect them looking backward." Thus, today's fantasy on IoT will be a reality of tomorrow.

Human inventions started from simpler ones like the wheel Archimedes screw, to many complex ones like telephones, cameras, smartphones, automobiles, flying machines, etc. All these inventions have made their position in society with massive changes in all domains from political to social. These inventions have settled in people's daily lives so profoundly that it makes it impossible for us to appreciate their importance (similar to how we appreciate the eyes or hands). Looking back at the past, we can say, not many people imagined a device like a smartphone which could also be used as a remote to control cars, helps seek answers, and who knows, might be a nuclear detonator also. These inventions have rewired the whole civilization and have reset our way of life.

Joe M. Bohlen, along with George M. Beal and Everett M. Rogers in 1957, presented the notion known as the technology adoption curve, as shown in Figure 1.1. They formulated the concept that every invention/solution/product follows a predictable path that can be traced back to the classical normal distribution, also known as the bell curve [6]. The distribution consists of innovators followed by early adopters to laggards. The formulated model still holds, but the period has accelerated at lightning speed, from century to decades, to years, and now into months. Internet of things

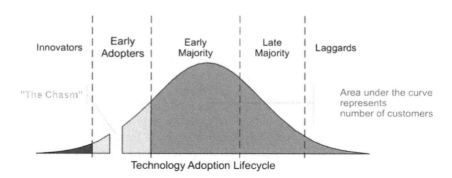

FIGURE 1.1 Technology adoption curve.

[7] is hanging between innovators and early adopters. In the future, it will be the backbone of life and businesses. Connected devices have existed since the introduction of computer networks and consumer electronics. But it wasn't until the Internet was introduced that researchers worldwide started to theorize a well-connected network of devices spread all around the earth with which humans can communicate and interact. A couple of years back, this fantasy started to become a reality in labs that now can be seen in everyday households; we call it the Internet of things.

Experts suggest that IoT is the second wave of the digital revolution [8,9] after computers and the Internet in the 1980s and 1990s, respectively. It is really hard to find the exact moment which led to this wave of digital revolution, but trend markers indicate Apple's iPhone introduction in 2007 starting a domino effect. It gave enormous power in the hands of the user with point-to-point communication and digital applications.

Now, smartphones consist of multiple sensors from microphone, camera, motion sensors, GPS, and many more, along with high computation power. These smartphones can serve as a dashboard [10,11] and/or remote control for IoT devices. Radio-frequency identification (RFID) technology has also come a long way [12–15], with which we can tag or track objects along with cloud computation which can serve as a door to the future. We generate more than 2.5 Quintillion bytes of data each day, and the rate of data generation is only increasing with time. With IoT in place, data collection will take the next steps which have never been explored before, from tracking animals, insects, birds, or viruses and finding patterns in nature which humans could not understand before due to its complexity. It is like finding a new dimension with all previous physics rules, which makes it open to exploration.

IoT can help us understand intricate patterns [16] or real-time [17] conditions of bridges, roadways, buildings, or super-efficient tracking systems, where the sky is the limit. The generation of new data from sensors has unexplored usage. This will allow big data and machine learning to play an important role in solving complex problems that used to be impossible earlier.

Today machine learning [18] has already settled in our life from Internet search to music recommendations. It has crossed the early adopter's mark in technological life cycle couple of years back, but with the combination of IoT and machine learning, it is still at the innovator's stage. However, the next phase won't be in years but rather months, after breaking the necessary data collection threshold. The race to collect IoT data [19–23] has already begun with smartphones, smartwatches, and smart speakers. New IoT devices are being theorized from old devices every day. The race to make smart electronics for the user has already begun, but the race path is still being explored. Devices such as smart bulbs, smart washing machines, or smart cars are just the tip of the iceberg in the world of IoT. The future of IoT with machine learning will change education [92,94], business, personal life, and politics [26] forever, probably for good. All technologies are dangerous in the wrong hands, and the amalgam of IoT with machine learning is no different [24,25].

The general rule in any revolution is that some people win and some people lose. IoT will make winners and open doors for new opportunities for them. But unfortunately, it will make many jobs of the current-era completely obsolete. Any potential gain of opportunity comes with plenty of challenges in IoT. A severe threat to privacy [27,28] hangs in danger, if the right actions in due time are not taken. Privacy-concerned potential weapons [29,30] for warfare, using the said technology, can be more dangerous than anything before. It won't be a surprise to know that development of such applications has already started in the background.

In this chapter, we have discussed the future, present, and past of Digital Twin Technology. IoT itself has immense potential to change the world, but the combination of IoT with machine learning has far more potential than anything else. The applications, technologies, and security of IoT products that deal with massive data are discussed thoroughly. All the machine learning models have very specific usage, and hence, their place in IoT systems is also essential. We have discussed all machine models used in IoT and the different methods to transfer data with their appropriate application in the chapter, along with IoT security infrastructure.

1.2 DIGITAL TWIN TECHNOLOGY

Machine learning has started to take its place in IoT systems [31], but due to certain limitations, its progress is slow. IoT data for machine learning is collected using two different methods, namely, stream and accumulated [32]. Stream data [33] is captured or generated in small intervals, whereas accumulated data is sent in a block for analysis, storage, or prediction. These two types of data [34] are different at the core; hence, their usage is also different. Response time in many applications has to be in real-time or near real-time [35,36]; hence, accumulated data won't be useful in such circumstances. Accumulated data is used where analytical data can be used even after certain days or hours after generation. For example, a self-driving car system cannot rely on accumulated data. It has to be based on streaming prediction, which can respond in few milliseconds to a few seconds. On the other hand, in recommendation systems, video analytic, climate forecasting, medical diagnosis, and finance, such areas can use accumulated data for analysis.

IoT streaming data for machine learning is usually deployed on the basis of high-performance cloud computers [37], data parallelism and incremental processing framework, used extensively [38]. Data parallelism [39] is performed by dividing extensive dataset into smaller datasets, on which analytics can be performed simultaneously. On the contrary, in incremental processing, a small batch of data is fetched and processed in a computational pipeline as quickly as possible. However, these techniques reduce time latency but are not the best solution for real-time IoT applications. Researches are trying to bring streaming data analytics close to the source (IoT device). The closer the computation/analytics to the source, less sensible it becomes to use parallelism and incremental processing. But bringing fast analytics close to the source introduces new challenges such as computation limitation, lesser storage, and fewer power resources. Deep learning is a subset of machine learning, consisting of supervised and unsupervised algorithms to train systems [40]. These algorithms are further categorized in specific architectures, and all such architectures consist of multiple layers. Each layer is responsible for producing a nonlinear response based on the input from the previous layer. Deep learning systems closely resemble the human brain to solve complex problems.

Deep learning (DL) has several advantages over traditional artificial neural networks (ANN); one of DL's significant advantages over ANN is learning hidden features using hidden layers. DL models are categorized into generative, discriminative, and hybrid. Discriminative models are usually based on supervised learning. However, generative models are based on unsupervised learning. In contrast, hybrid models are the best of both worlds, discriminative and generative.

DL architectures include many architectures such as convolutional neural networks (CNNs), recurrent neural networks (RNNs), long short-term memory (LTSM), autoencoders (AEs), variational autoencoders (VAEs), generative adversarial network (GANs), and restricted Boltzmann machine (RBMs). All these architectures perform an extremely specific function in IoT systems.

1.3 DIGITAL TWIN TECHNOLOGY USING CONVOLUTIONAL NEURAL NETWORKS WITH IoT

Convolutional neural networks (CNN) are well known for processing pixel data and works wonders with image recognition. The conventionally vision-based task in DL failed to learn features that might transform, such as rotation, translation, or reflection, but CNN overcame these issues using hidden layers. Hidden layers are a combination of convolution layers and are fully connected end layers. The convolutional layer is at the core of a CNN. The benefit of CNN in IoT [41] is that many IoT devices use cameras as a sensor to sense the environment around them. Products that majorly depends upon cameras are drones, smart connected cars, and smartphones.

1.4 DIGITAL TWIN TECHNOLOGY USING RECURRENT NEURAL NETWORKS WITH IoT

Recurrent neural networks (RNN) are the most useful DL architecture in IoT. RNN works flawlessly in real-time hence streaming data analytics in many major devices depends upon RNN. RNN is majorly known for its property to predict the outcome, based on several previous samples, making it a perfect candidate to analyze the sequence of inputs and rate of change in many streaming data. RNN use Backpropagation Through Time (BPTT) to train the network. RNN is used in those applications where individual samples are not enough, but sequences of inputs play a vital role. The structure of RNN is shown in Figure 1.2.

Energy demand prediction for smart home in smart grid framework [42] is one of the products based on RNN with IoT. Other products are intrusion detection system [43] and breathing-based authentication on resource-constrained devices [44].

1.5 DIGITAL TWIN TECHNOLOGY USING LONG SHORT-TERM MEMORY WITH IoT

Long short-term memory (LSTM) is an extension of RNN [45]. LSTM consists of the memory cell, as the drawback in RNN is that it forgets the past. This constraint has been overcome by introducing gates' concept. There are three types of gates in

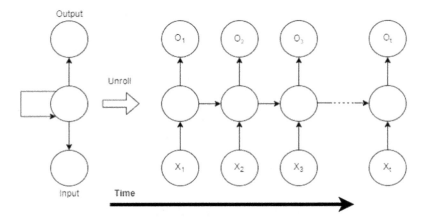

FIGURE 1.2   Structure of RNN.

a memory cell, forget gate, read gate, and write gate. All the gates work together to perform multiple functions from write, read, and delete. One of the significant differences between LSTMs and RNNs is forgetting gate, which ensures cell states. Long dependency in time on data makes LSTM perform better than RNN. This long dependency makes the IoT data a perfect candidate for prediction. Applications such as human activity recognition in sports, disaster prediction on environmental monitoring, and academic performance predictions are some of such applications that use this concept.

LSTM has many industrial IoT applications which produce a huge amount of data [46]. Water quality prediction [47], QoS attributes forecasting [48], and PM10 concentration prediction [49] are also based on LSTM concept.

1.6   DIGITAL TWIN TECHNOLOGY USING AUTOENCODERS AND VARIATIONAL AUTOENCODERS WITH IoT

Autoencoders (AEs) have an input layer and an output layer which are connected thought multiple hidden layers that can vary from one to tens. AEs are also known to have a same number of input and output layers as shown in Figure 1.3. The task of AEs is to convert the input to output without changing input very much. Such networks are used mainly for unsupervised learning and transfer learning [50]. The major uses of AEs are in fault detection, medical diagnosis, and anomaly detection in the assembly line performance. Autoencoder application consists of IoT botnet detection [51] and distributed anomaly detection in WSN for IoT [52].

Later, autoencoders were modified into variational autoencoders (VAEs) in 2013. VAE uses a fast-training process by using backpropagation [53]; this model is usually used for semi-supervised learning [54]. IoT's diverse data and its scarcity of labeled data make VAE a great fit for IoT solutions. VAE is mostly used in security systems and failure detection. The principal behind VAE relies on two networks: one generating samples, while the other performing approximate inference. VAE is usually used

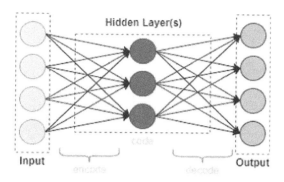

FIGURE 1.3   Autoencoder network structure.

in the smart city [55] and IoT cybersecurity [56]. VAE provides an advantage in cases of smart cities as there is an extreme scarcity of labeled data, even though a large number of sensors generate huge data, but no class labeled is obtained. In such areas, the VAE IoT system plays a vital role [55].

1.7   DIGITAL TWIN TECHNOLOGY USING GENERATIVE ADVERSARIAL NETWORKS WITH IoT

Generative adversarial networks (GANs) [57] were introduced recently in 2014 by Ian Goodfellow and his colleagues. GAN consists of two neural networks, namely, generative and discriminative networks, as shown in Figure 1.4. The generative network is responsible for generating new data using the available dataset after it learns data distribution. At the same time, discriminative networks try to differentiate between real and fake data. The generative network works against a discriminative network to try to deceive it, while discriminative networks try to tell the difference between fake and real. Both the networks in GAN compete against each other in a minimax game; one network tries to maximize the difference between real and fake, while the other tries to minimize it. In IoT, GANs can be used as applications that require data creation from the available dataset. For example, localization and path-finder, [58] in this former network, generate multiple paths between two points, whereas in the latter, they try to identify the most viable path.

1.8   DIGITAL TWIN TECHNOLOGY USING RESTRICTED BOLTZMANN MACHINE WITH IoT

Restricted Boltzmann machines (RBMs) is one of the oldest artificial neural networks generated in 1986 by Paul Smolensky [59]. Today after several modifications, RBMs are used in classification, collaborative filtering, feature learning, and topic modeling. RBM is a two-layer model, where one is the visible layer, which we know as input, and the other is the hidden layer responsible for latent variables. In IoT, RBMs are used in multiple verticals, from energy consumption prediction, traffic congestion prediction, posture analysis, to indoor localization. In other words, all the

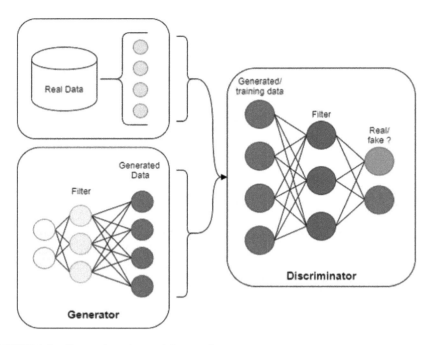

FIGURE 1.4   Generative adversarial network structure.

applications rely upon extracting the essential features from the bucket of available features [60,61].

1.9   DEEP LEARNING ON IoT ALONG WITH OTHER ALGORITHMS

Use of conventional DL with IoT has been all time high since 2015; however, use of other technologies along with DL is also marking their position in the world. Deep reinforcement learning (DRL) is fusion of reinforcement learning (RL) with DNNs. In IoT, DRL is used for semi-supervised training to perform localization in campus [62]. Transfer learning and online learning [63] are some more approaches that are combined with conventional DL to get better results. As of now, transfer learning with DL on IoT is an active research topic, and much of its work is still under controlled labs which have not been used in the real world.

1.9.1   DIGITAL TWIN TECHNOLOGY WITH IoT APPLICATIONS

Fundamental services in IoT that depends upon DL are signal processing, natural language processing, and image recognition [64]. These fundamental services further consist of several applications such as oil pipeline monitoring [65,66], smart speakers [67], energy monitoring, security systems [68], and indoor localization [69]. Indoor localization [70,71] is used in smart homes, smart campuses, and hospitals.

1.9.2 Digital Twin Technology with Smart Homes

Smart homes [72] are combination of a huge variety of products based on IoT technology used to enhance homes. These products consist of smart energy meters [73], smart water meters, and smart devices (light bulbs, refrigerator, air conditioner, heaters, washing machine, and geyser). All these smart devices can be controlled using smartphones as a remote or smart speakers such as Alexa [74] or Google home. One fourth of electricity in India is used for household activities [75]. This demand will only grow in coming years; therefore, it is imperative to improve energy efficiency and predict future energy requirement for smart homes. Electricity load prediction today is an important field of research for all IoT enthusiasts. It will allow us to create bidirectional data flow for smart electricity grid which will allow us to predict the demand of the future in advance and generate electricity accordingly.

1.9.3 Digital Twin Technology with Smart City

Similar to smart homes, smart cities are combination of a wide range of IoT products [76]. Many domains such as transportation [77], energy, agriculture [78], etc. are part of smart city infrastructure. Smart cities are the hottest topic among machine learning enthusiasts rather than IoT [79]. Machine learning in smart cities can help in crowd movement patterns analysis, security systems, and public transport. Waste in today's world is one of the biggest problems; hence, waste management system using IoT sensors in smart cities is the most researched topic. The most straightforward built waste management system is using computer vision along with CNN. Air quality in India is also a major concern. In 2020, CNN (news agency) reported that out of the 30 most polluted cities in the world, 21 are in India. Air index monitoring and prediction of pollution is another major research in IoT smart city application.

With ever-increasing automobile industry with limited land, parking vehicles has become an all-time-high problem. Due to unauthorized parking, traffic jams and accidents are increasing rapidly. Therefore, IoT products can help to manage parking in smart cities [80].

1.9.4 Digital Twin Technology with Energy

Energy demand is an ever-increasing industry in modern world. It is believed that after fire, if anything that has changed the future forever is the discovery of electricity. We have come a long way in understanding electricity, but since the last couple of decades, not much has changed in electric grid systems. All electric grid systems work on unidirectional data flow, and therefore it is high time in 2020 to use IoT, to make data flow bidirectionally. Smart grids [81], renewable energy generation (solar and wind), and smart meters are currently active fields of research. DL has been used extensively in energy domain. Solar power prediction takes weather as input with numeric value for weather [82]. Due to these reasons, Auto-LSTM (combination of AEs and LSTMs) have shown the best results from all other DL algorithms. Auto-LSTM have outperformed all other algorithms as Auto-LSTM are able to extract features from the raw data, while others fail to do so.

1.9.5 DIGITAL TWIN TECHNOLOGY USING INTELLIGENT TRANSPORTATION SYSTEM

Data generated from intelligent transport system (ITS) is increasing every day. Intelligent transport system is a key component in the smart city infrastructure. It aims to reduce traffic and provide an efficient way to commute. Current research [a1] has allowed us to process data on the vehicle, before being transmitted to the cloud. Self-driving cars are a major part of intelligent transportation systems. Self-driving cars [83] use extensive DNN algorithms to perform multiple tasks, from detecting pedestrians, analyzing traffic signals and detecting objects along with identifying other cars on the road.

ITS recommended by Thiyagarajan and his team consists of three components: a sensor, a monitoring system, and a display system. A sensor consists of global positioning system (GPS), near field communication (NFC), and temperature and humidity sensors. These sensors are connected to Internet via GSM network, and all the data in raw form is sent to monitoring system to generate usable data.

1.9.6 DIGITAL TWIN TECHNOLOGY IN HEALTHCARE AND LIFESTYLE

IoT patched up with machine learning is used to provide better lifestyle and health-care to communities [84]. Smart watches and rings help user to measure heart beat rate, blood oxygen saturation, blood pressure, electrocardiogram, and body tempera-ture. IoT used for disease detection in humans using machine learning has been an active field in both academic and industry.

Frameworks like mutual authentication [85] for session key distribution system for devices with encryption modules. Many researchers use those self-quantification that use IoT wearable devices to detect different anomalies in one's health. These self-quantified devices allow us to monitor vitals in human body and send alerts or notification if a problem is detected.

1.9.7 DIGITAL TWIN TECHNOLOGY IN AGRICULTURE

With increasing population demand for food at an all-time high, we can only expect the demand to increase in future. Therefore, the need to produce healthy crops with efficient ways is the need of the hour. Computer vision along with machine learn-ing has shown positive results in disease recognition. Places in India, South Africa, and Zambia face acute water problems, transport or logistics, in which case IoT has shown promising results [87] to solve these problems.

Low-power wireless sensors, characteristics verification, and machine learning models are widely used to verify quality of the production [88]. Grape leaf [89], cot-ton [89], and tomato disease detections, using machine learning models, are some of the common machine learning usages [90]. Green IoT, which uses agriculture and healthcare applications (GAHA), has its own architecture and uses in today's world [91]. The most popular design in green IoT is sensor-cloud integration model. GAHA components include networks such as HAN, WPAN, or WSN. However, RFID plays an important role in green IoT used by GAHA. Therefore, use of green

RFID was an important research. It uses smaller tags along with decreasing amount of non-degradable material.

1.9.8 Digital Twin Technology in Education

Education system in India and other third-world countries needs serious upgradation. IoT along with machine learning has shown viable options to achieve such upgrades [92]. Mobile devices, laptops, tablets, and computers via web applications or mobile applications allow us to gather information on the user to perform deep learning and how analytical methods are used to understand his/her needs along with keeping record on their progress and achievements.

Attention span of normal human being is in constant decline in technology-centric world. However, in 2000, attention span was 12 seconds, but it has reduced to 8 seconds today. To get perspective of things, the attention span of gold fish is 8 seconds. Due to decrease in attention span, general way to consume knowledge through blogs, books, or methods developed in the last couple of years' podcasts or videos could become history in a decade. Virtual reality (VR) [93] and augmented reality (AR) [94] are the future of education, and their potential use with IoT devices and machine learning analytics can open door to the future.

1.9.9 Digital Twin Technology in Industry

Industries require fast and high precision systems. Therefore, IoT and cyber-physical systems (CPS) have become core-elements of the manufacturing industry. CPS systems are being used regularly but IoT systems are still far from being deployed. Use of IoT and CPS together, along with ML, is expected to revolutionize industry which is known as industry 4.0. Computer vision using DL (CNN) is playing a vital role in industry 4.0. A wide variety of objects from quality control, to fault diagnosis and product line inspection are parts of it. Further logistics using RFID and beacons will play important roles for products to be delivered from manufacturer to user. Major architectures used CNN, AlexNet, and GoogleNet for visual inspection.

1.9.10 Digital Twin Technology in Government

All inventions have both technological and sociological effects, IoT is no different. IoT with machine learning has great potential advantages for intelligent and enhanced connectivity. Prediction of natural disaster (floods, hurricane, forest fires, etc.) along with environmental monitoring for important areas is of high importance, and so the governments should take required action in due time [95].

Along with environmental prediction, several other usages of this technology are infrastructure damage detection, like roads, water pipelines, oil pipelines, and important defense routes. Road damage detection using DNNs, with the help of crowd sourcing data, using IoT enabled devices has been presented in Ref. [96]. Data analysis is not possible on mobile IoT devices; hence, all the data has to be sent to the cloud server for training and analysis. The report generated from the cloud server is used further to fix the infrastructure.

1.9.11 DIGITAL TWIN TECHNOLOGY IN SPORTS

Sports is a vital part of any civil society and since introduction of Internet it has changed extensively. Recent analytical studies on professional sports player have allowed the management to form high-performance teams ** ADD Citation from main paper **. Methods to produce smart sports arena have been in increasing demand. Some of such ideas are presented in [Citation from main paper], that are used in deep learning to design intelligent basketball arena.

Several IoT products have been developed or modified to track body stats of the player to understand it's requirement when player is in the field. Smart devices such as shoes, bands, watches or clothes are researched extensively for the future product development.

1.9.12 DIGITAL TWIN TECHNOLOGY IN RETAIL

E-commerce has transformed retail years beyond imagination. Introduction of IoT in retail will have similar impact. RFID-based products, such as smart carts, computer vision-based stores [97], and several other products, are some of the examples. Recommending shoppers about the items/products is also an interesting application of IoT technologies. Several methods have been proposed to analyze customer interest in the merchandise [98]. Architectures used in retails include CNN, RNN, and DNN.

1.9.13 DIGITAL TWIN TECHNOLOGY IN SMART IoT INFRASTRUCTURE

IoT infrastructure is an amalgam of sensors, actuators, media, and several other devices which generate M2M traffic 24 * 7. The expected M2M traffic in the near future is only expected to increase; however, at current state, these data packets may not congest the backbone of the Internet, but with an increase in demand, such congestion can take place. Also, even at the current rate of transmission, some delay in data transmission has been noticed in channel access phase. To solve such problem, load balancing uses DL architecture, as proposed by Kim [99], which is a viable solution.

However, current introduction to 5G network infrastructure has opened a vast number of doors for the IoT devices. It is believed that crowd-sourced cellular networks along with big data will allow us create more reliable routes and high-precision coverage to improve data transmission performance.

1.9.14 DIGITAL TWIN TECHNOLOGY IN ON-BOARD ML ON IoT DEVICES

Most of the research on ML has taken place on cloud servers. These servers have huge storage and high computation power to run ML algorithms. But with emergence of IoT devices, storage and computation power have become one of the biggest constrains. However, these devices also need real-time analytics and resources are constrained; therefore, research on resource efficient algorithm has started. It has been noticed that a large number of hidden layers are not the key to high accuracy [100]. If useless parameters and layers are removed in a considerable manner, it will make these algorithms less complex [100,101] and more IoT-friendly. ML methods have millions to billions of parameters, and such parameters require huge computation power and extremely high storage to perform ML algorithms. However, certain methods and technologies can be used to decrease computation for ML on IoT devices.

1.9.15   DIGITAL TWIN TECHNOLOGY IN NETWORK COMPRESSION

Network compression is performed by converting dense networks to sparse network. Network compression reduces the storage and computational requirement to apply ML algorithms to perform classification and prediction on IoT devices. However, like any method, network compression also has its own limitations. It does not support all networks in general. It has a very specific list of networks which support network compression. In Ref. [102], a study was performed that allowed to adopt compressed ML models on IoT devices.

Researchers [103] made an inference engine known as EIE, which has a specific hardware and contains SRAMs instead of DRAMs that work great with compressed network models. In the designed architecture, sparse matrix weight and vector multiplication works without losing its efficiency. The energy usage has been declined to 120 lesser than normal models.

In [104], Courbariaux and his team designed a binarize network weights and neurons at training and inference phase which helps to reduce memory footprint and accesses. This network allows the system to perform arithmetic operations using bit-wise operations, which decreases power consumption up to a great extent.

1.9.16   DIGITAL TWIN TECHNOLOGY IN APPROXIMATE COMPUTING

In many machine learning applications, exact or absolute value is not required therefore approximate values solve the problem. These methods allow us to save both computation power and energy consumption of the IoT device [105, 106].

Researchers have used approximate computing in ML models for devices with limited resources. Venkataramani [105] and his team designed a model which uses approximate computing in neural networks to form approximate neural network.

Even though approximate computing solves problem of resource-constrained devices, training of the model takes place on resource extensive systems. Later, trained models are installed on resource-constrained devices for further use of model.

1.9.17   DIGITAL TWIN TECHNOLOGY IN HARDWARE IMPROVEMENT

While improvement in software is all time high in researchers, hardware improvement is equally important. Therefore, there has been active study in improving hardware and circuits which results in the energy efficient products with low memory usage [107,108].

1.9.18   DIGITAL TWIN TECHNOLOGY WITH ML FOR IoT INFRASTRUCTURE

Cloud computing was believed to be the solution for IoT to use ML analytics. However, cloud computation has its own drawbacks such as security issues to send

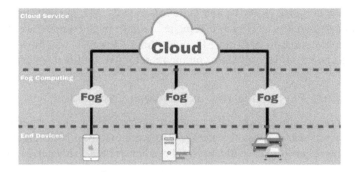

FIGURE 1.5   Fog computing infrastructure.

personal data over the network, legal constrains (some countries do not allow cloud servers to be hosted outside national borders) and undeniable time latency. In some cases, it is required to combine data from several IoT data sources to have resourceful prediction/analytics, but unfortunately, we do not have such computation available to perform analytics on individual nodes (IoT devices).

To solve the cloud computing problem, a new generation of infrastructure is proposed which allows data abstraction and computation closer to the device known as fog computing. The emergence of fog computing is based on the fact that the number of IoT devices is constantly increasing and such an increasing number of devices put strain on the cloud sever. However, these servers are not able to handle the data; therefore, fog computing was invented. Figure 1.5 shows fog computing infrastructure. The major difference between fog computing and cloud computing is that the former provides decentralized local access, whereas the latter provides centralized access.

Fog computing is recently proposed as it brings analytics and computing closer to the devices or end-users. Fog-based analytics allow us to use the benefits of cloud computing while surpassing its drawbacks such as network security and latency. Fog has allowed the IoT devices to perform real-time data analytics to some extent because the source is closer to the fog.

1.9.19   DIGITAL TWIN TECHNOLOGY IN IoT INFRASTRUCTURE SECURITY

IoT devices and infrastructures are not safe from security vulnerabilities. All the IoT devices are extremely prone to hackers, and they contain personal information of the user from health statistics to finger prints. At the same time, some devices are responsible for energy demand prediction and usage. False data injection (FDI) is a common attack on data driven systems. In these types of attacks, false data is entered into the system which makes the system predict false information and give unreliable results. IoT devices mostly consist of three layers, namely, perception layer, network layer, and application layer.

Perception layer has security concerns such as signal strength, hardware tampering, and other types of attacks. Perception layer's confidentiality can easily be exploited using repay attack, timing attack, or node capture attack. However, integrity of this layer is not safe from the attack. It can send malicious data by adding another node to the network by the attacker and can destroy network's integrity. Draining energy, using DoS attack, is also one of the major security flaws. All these attacks can be solved by adding point-to-point encryption and verifying the device before data processing but many active IoT devices failed to add most basic security features.

Network layer is not safe from DoS attack either, as apart from DoS attacks, man-in-the-middle, followed by eavesdropping is one of the major security flaws of the devices. Eavesdropping on IoT devices can result in identity theft which is one of the major crimes in the 21st century. IoT devices introduce a new way of communication called machine-to-machine communication which introduces compatibility as a new security issue. Heterogeneity of the network makes current security measures obsolete on many fronts, which opens new doors for the attackers to benefit from it. Research to stop network layer attacks is still ongoing.

Application layer in IoT does not have any global policy for application development; therefore, there are many security flaws. Different IoT devices consist of different authentication mechanisms, data transfer protocols, and designs. These differences make data privacy and identification of the authorized devices difficult.

1.10   DIGITAL TWIN TECHNOLOGY: FUTURE OF IoT

Future of IoT is bright and has limitless potential. The potential is not limited to enabling billions of devices simultaneously but rather using high volume of data which can be used to automate the businesses and perform analytics. It is estimated that by 2025, approximately 75.44 billion IoT devices will be active all across the globe. However, the annual growth rate of IoT from 2020 to 2025 is projected to be 28.7%. The estimated data generation from sensors by 2025 would increase to 79.4 zettabytes.

As an old saying goes, "With great power comes great responsibilities" and similarly, "with great data comes great analytics". IoT will generate tremendous amount of data and that will provide corporates and researchers to work on the same.

While benefits of IoT are limitless, so are the challenges. Currently we lack labeled IoT dataset and similar challenges are presented by prepossessing onboard IoT devices. While we have machine learning to solve big IoT data analytics problems, it still has some limitations. CNN algorithms have shown false predictions in many cases where machine learning predicted a familiar object, but humans denied its existence. Applications such as electricity load forecasting and temperature forecasting still need improvement at their analytic cores. IoT with machine learning will open doors that were earlier not even known to have existed.

REFERENCES

1. Hooper, P. and Richardson, J.D., 1991. International economic transactions: Issues in measurement and empirical research (No. w3805). National Bureau of Economic Research.
2. Böhme, R., Christin, N., Edelman, B. and Moore, T., 2015. Bitcoin: Economics, technology, and governance. *Journal of Economic Perspectives*, 29(2), pp.213–238.

3. Ivatury, G., 2009. Using technology to build inclusive financial systems. In Pischke J., Matthäus-Maier I. (eds.), *New Partnerships for Innovation in Microfinance*. Springer, Berlin, Heidelberg (pp. 147–172). https://doi.org/10.1007/978-3-540-76641-4_9.

4. Mujawar, A., Krishnan, S.S., Kumar, S. and Sawant, A., 2020. "IoT: Green Data Center Strategies."

5. Gupta, P., Dedeoglu, V., Najeebullah, K., Kanhere, S.S. and Jurdak, R., 2020. Energy-aware Demand Selection and Allocation for Real-time IoT Data Trading. arXiv preprint arXiv:2002.02074.

6. Rogers, E. M., Singhal, A., and Quinlan, M. M., 2014. Diffusion of innovations (pp. 432–448). Routledge. New York.

7. Atzori, L., Iera, A. and Morabito, G., 2010. The Internet of things: A survey. *Computer Networks*, 54(15), pp.2787–2805.

8. Merritt, B., 2016. The digital revolution. *Synthesis Lectures on Emerging Engineering Technologies*, 2(4), pp.1–109.

9. Roman, D.H., Conlee, K.D., Abbott, I., Jones, R.P., Noble, A., Rich, N., Ro, I., Kaufman, J., Weikert, R. and Costa, D., 2015. *The Digital Revolution Comes to US Healthcare*. New York: Goldman Sachs.

10. Wang, D., Lo, D., Bhimani, J. and Sugiura, K., 2015, July. Anycontrol–IoT based home appliances monitoring and controlling. In *2015 IEEE 39th Annual Computer Software and Applications Conference* (Vol. 3, pp. 487–492). IEEE.

11. Aloi, G., Caliciuri, G., Fortino, G., Gravina, R., Pace, P., Russo, W. and Savaglio, C., 2017. Enabling IoT interoperability through opportunistic smartphone-based mobile gateways. *Journal of Network and Computer Applications*, 81, pp.74–84.

12. Raza, N., Bradshaw, V. and Hague, M., 1999. Applications of RFID technology. 1–1. IEE. London.

13. Kumar, S., Swanson, E. and Tran, T., 2009. RFID in the healthcare supply chain: Usage and application. *International Journal of Health Care Quality Assurance*,22(1), pp.67–81.

14. Gupta, P. and Margam, M., 2017. RFID technology in libraries: A review of literature of Indian perspective. *DESIDOC Journal of Library Information Technology*, 37(1), p.58.

15. Zhu, X., Mukhopadhyay, S.K. and Kurata, H., 2012. A review of RFID technology and its managerial applications in different industries. *Journal of Engineering and Technology Management*, 29(1), pp.152–167.

16. Akbar, A., Carrez, F., Moessner, K. and Zoha, A., 2015, December. Predicting complex events for pro-active IoT applications. In *2015 IEEE 2nd World Forum on Internet of Things (WF-IoT)* (pp. 327–332). IEEE.

17. Akbar, A., Khan, A., Carrez, F. and Moessner, K., 2017. Predictive analytics for complex IoT data streams. *IEEE Internet of Things Journal*, 4(5), pp.1571–1582.

18. Alpaydin, E., 2020. *Introduction to Machine Learning*. MIT Press.

19. Plageras, A.P., Psannis, K.E., Stergiou, C., Wang, H. and Gupta, B.B., 2018. Efficient IoT-based sensor BIG Data collection–processing and analysis in smart buildings. *Future Generation Computer Systems*, 82, pp.349–357.

20. Orsino, A., Araniti, G., Militano, L., Alonso-Zarate, J., Molinaro, A. and Iera, A., 2016. Energy efficient IoT data collection in smart cities exploiting D2D communications. *Sensors*, 16(6), p.836.

21. Luo, E., Bhuiyan, M.Z.A., Wang, G., Rahman, M.A., Wu, J. and Atiquzzaman, M., 2018. Privacyprotector: Privacy-protected patient data collection in IoT-based healthcare systems. *IEEE Communications Magazine*, 56(2), pp.163–168.

22. Luong, N.C., Hoang, D.T., Wang, P., Niyato, D., Kim, D.I. and Han, Z., 2016. Data collection and wireless communication in Internet of Things (IoT) using economic analysis and pricing models: A survey. *IEEE Communications Surveys Tutorials*, 18(4), pp.2546–2590.

23. Wang, W., Xu, P. and Yang, L.T., 2018. Secure data collection, storage and access in cloud- assisted IoT. *IEEE Cloud Computing*, 5(4), pp.77–88.
24. Abbasy, M.B. and Quesada, E.V., 2017. Predictable influence of IoT (Internet of Things) in the higher education. *International Journal of Information and Education Technology*, 7(12), pp.914–920.
25. Mrabet, H.E. and Moussa, A.A., 2017. Smart classroom environment via IoT in basic and secondary education. *Transactions on Machine Learning and Artificial Intelligence*, 5(4). http://dx.doi.org/10.14738/tmlai.54.3191
26. Monteiro, E. and Parmiggiani, E., 2019. Synthetic knowing: The politics of the Internet of things. arXiv preprint arXiv:1903.00663.
27. Amyx, S., 2017. Privacy dangers of wearables and the Internet of things. In Andrew Marrington, Don Kerr and John Gammack (eds.), *Managing Security Issues and the Hidden Dangers of Wearable Technologies* (pp. 131–160). IGI Global.
28. Beale, S.S. and Berris, P., 2017. Hacking the Internet of Things: Vulnerabilities, dangers, and legal responses. *Duke Law Technology Review*, 16, p.161.
29. Cha, S., Baek, S., Kang, S. and Kim, S., 2018. Security evaluation framework for military IOT devices. *Security and Communication Networks*, 2018, p. 12. doi: 10.1155/2018/6135845.
30. Yushi, L., Fei, J. and Hui, Y., 2012, May. Study on application modes of military Internet of Things (MIOT). In *2012 IEEE International Conference on Computer Science and Automation Engineering (CSAE)* (Vol. 3, pp. 630–634). IEEE.
31. Tang, J., Sun, D., Liu, S. and Gaudiot, J.L., 2017. Enabling deep learning on IoT devices. *Computer*, 50(10), pp.92–96.
32. Ahmed, E., Yaqoob, I., Hashem, I.A.T., Khan, I., Ahmed, A.I.A., Imran, M. and Vasilakos, A.V., 2017. The role of big data analytics in Internet of Things. *Computer Networks*, 129, pp.459–471.
33. Chen, C.Y., Fu, J.H., Sung, T., Wang, P.F., Jou, E. and Feng, M.W., 2014, August. Complex event processing for the Internet of things and its applications. In *2014 IEEE International Conference on Automation Science and Engineering (CASE)* (pp. 1144–1149). IEEE.
34. Mohammadi, M., Al-Fuqaha, A., Sorour, S. and Guizani, M., 2018. Deep learning for IoT big data and streaming analytics: A survey. *IEEE Communications Surveys Tutorials*, 20(4), pp.2923–2960.
35. Verma, S., Kawamoto, Y., Fadlullah, Z.M., Nishiyama, H. and Kato, N., 2017. A survey on network methodologies for real-time analytics of massive IoT data and open research issues. *IEEE Communications Surveys & Tutorials*, 19(3), pp.1457–1477.
36. Qu, T., Lei, S.P., Wang, Z.Z., Nie, D.X., Chen, X. and Huang, G.Q., 2016. IoT-based real-time production logistics synchronization system under smart cloud manufacturing. *The International Journal of Advanced Manufacturing Technology*, 84(1–4), pp.147–164.
37. Cai, H., Xu, B., Jiang, L. and Vasilakos, A.V., 2016. IoT-based big data storage systems in cloud computing: Perspectives and challenges. *IEEE Internet of Things Journal*, 4(1), pp.75–87.
38. Li, B., Diao, Y. and Shenoy, P., 2015. Supporting scalable analytics with latency constraints. *Proceedings of the VLDB Endowment*, 8(11), pp.1166–1177.
39. Shallue, C.J., Lee, J., Antognini, J., Sohl-Dickstein, J., Frostig, R. and Dahl, G.E., 2018. Measuring the effects of data parallelism on neural network training. arXiv preprint arXiv:1811.03600.
40. Shrestha, A. and Mahmood, A., 2019. Review of deep learning algorithms and architectures. *IEEE Access*, 7, pp.53040–53065.
41. Motamedi, M., Fong, D. and Ghiasi, S., 2016. Fast and energy-efficient CNN inference on IoT devices. arXiv preprint arXiv:1611.07151.

42. Munir, M.S., Abedin, S.F., Alam, M.G.R. and Hong, C.S., 2017. RNN based energy demand prediction for smart-home in smart-grid framework, pp. 437–439.
43. Larijani, H., Ahmad, J. and Mtetwa, N., 2019, July. A heuristic intrusion detection system for Internet-of-Things (IoT). *In Intelligent Computing-Proceedings of the Computing Conference* (pp. 86–98). Springer, Cham.
44. Chauhan, J., Seneviratne, S., Hu, Y., Misra, A., Seneviratne, A. and Lee, Y., 2018. Breathing-based authentication on resource-constrained IoT devices using recurrent neural networks. *Computer*, 51(5), pp.60–67.
45. Sherstinsky, A., 2020. Fundamentals of recurrent neural network (rnn) and long short-term memory (lstm) network. *Physica D: Nonlinear Phenomena*, 404, p.132306.
46. Zhang, W., Guo, W., Liu, X., Liu, Y., Zhou, J., Li, B., Lu, Q. and Yang, S., 2018. LSTM-based analysis of industrial IoT equipment. *IEEE Access*, 6, pp.23551–23560.
47. Liu, P., Wang, J., Sangaiah, A.K., Xie, Y. and Yin, X., 2019. Analysis and prediction of water quality using LSTM deep neural networks in IoT environment. *Sustainability*, 11(7), p.2058.
48. White, G., Palade, A. and Clarke, S., 2018, July. Forecasting QoS attributes using LSTM networks. In *2018 International Joint Conference on Neural Networks (IJCNN)* (pp. 1–8). IEEE.
49. Kim, S.K. and Oh, T.I., 2018. Real-time PM10 concentration prediction LSTM model based on IoT streaming sensor data. *Journal of the Korea Academia-Industrial Cooperation Society*, 19(11), pp.310–318.
50. Baldi, P., 2012, June. Autoencoders, unsupervised learning, and deep architectures. In *Proceedings of ICML Workshop on Unsupervised and Transfer Learning* (pp. 37–49).
51. Meidan, Y., Bohadana, M., Mathov, Y., Mirsky, Y., Shabtai, A., Breitenbacher, D. and Elovici, Y., 2018. N-BaIoT—Network-based detection of IoT botnet attacks using deep autoencoders. *IEEE Pervasive Computing*, 17(3), pp.12–22.
52. Luo, T. and Nagarajan, S.G., 2018, May. Distributed anomaly detection using auto-encoder neural networks in WSN for IoT. In *2018 IEEE International Conference on Communications (ICC)* (pp. 1–6). IEEE.
53. Doersch, C., 2016. Tutorial on variational autoencoders. arXiv preprint arXiv:1606.05908.
54. Kingma, D.P., Mohamed, S., Rezende, D.J. and Welling, M., 2014. Semi-supervised learning with deep generative models. In *Advances in Neural Information Processing Systems* (pp. 3581–3589).
55. Mohammadi, M., Al-Fuqaha, A., Guizani, M. and Oh, J.S., 2017. Semisupervised deep reinforcement learning in support of IoT and smart city services. *IEEE Internet of Things Journal*, 5(2), pp.624–635.
56. Yousefi-Azar, M., Varadharajan, V., Hamey, L. and Tupakula, U., 2017, May. Autoencoder-based feature learning for cyber security applications. In *2017 International Joint Conference on Neural Networks (IJCNN)* (pp. 3854–3861). IEEE.
57. Goodfellow, I., Pouget-Abadie, J., Mirza, M., Xu, B., Warde-Farley, D., Ozair, S., Courville, A. and Bengio, Y., 2014. Generative adversarial nets. In *Advances in Neural Information Processing Systems* (pp. 2672–2680).
58. Mohammadi, M., Al-Fuqaha, A. and Oh, J.S., 2018, July. Path planning in support of smart mobility applications using generative adversarial networks. In *2018 IEEE International Conference on Internet of Things (iThings) and IEEE Green Computing and Communications (GreenCom) and IEEE Cyber, Physical and Social Computing (CPSCom) and IEEE Smart Data (SmartData)* (pp. 878–885). IEEE.
59. Fischer, A. and Igel, C., 2012, September. An introduction to restricted Boltzmann machines. In *Iberoamerican Congress on Pattern Recognition* (pp. 14–36). Springer, Berlin, Heidelberg.

60. Elsaeidy, A., Munasinghe, K.S., Sharma, D. and Jamalipour, A., 2019. Intrusion detection in smart cities using Restricted Boltzmann Machines. *Journal of Network and Computer Applications*, 135, pp.76–83.
61. Sun, X., Ma, S., Li, Y., Wang, D., Li, Z., Wang, N. and Gui, G., 2019. Enhanced echo-state Restricted Boltzmann Machines for network traffic prediction. *IEEE Internet of Things Journal*, 7(2), pp.1287–1297.
62. Mohammadi, M., Al-Fuqaha, A., Guizani, M. and Oh, J.S., 2017. Semisupervised deep reinforcement learning in support of IoT and smart city services. *IEEE Internet of Things Journal*, 5(2), pp.624–635.
63. Grotov, A. and de Rijke, M., 2016, July. Online learning to rank for information retrieval: Sigir 2016 tutorial. In *Proceedings of the 39th International ACM SIGIR Conference on Research and Development in Information Retrieval* (pp. 1215–1218).
64. Soldatos, J. ed., 2016. *Building Blocks for IoT Analytics*. River Publishers.
65. Sun, J., Zhang, Z. and Sun, X., 2016. The intelligent crude oil anti-theft system based on IoT under different scenarios. *Procedia Computer Science*, 96, pp.1581–1588.
66. Priyanka, E.B., Maheswari, C. and Thangavel, S., 2018, December. Proactive decision making based IoT framework for an oil pipeline transportation system. In *International Conference on Computer Networks, Big Data and IoT* (pp. 108–119). Springer, Cham.
67. Koo, H., Kim, S. and Nam, C., 2017. Speaker Wars begins: Which applications will be the killer content for smart speaker?, 14th Asia-Pacific Regional Conference of the International Telecommunications Society (ITS): "Mapping ICT into Transformation for the Next Information Society", Kyoto, Japan, 24th–27th June, 2017, International Telecommunications Society (ITS), Calgary.
68. Xiao, L., Wan, X., Lu, X., Zhang, Y. and Wu, D., 2018. IoT security techniques based on machine learning: How do IoT devices use AI to enhance security? *IEEE Signal Processing Magazine*, 35(5), pp.41–49.
69. BniLam, N., Ergeerts, G., Subotic, D., Steckel, J. and Weyn, M., 2017, September. Adaptive probabilistic model using angle of arrival estimation for IoT indoor localization. In *2017 International Conference on Indoor Positioning and Indoor Navigation (IPIN)* (pp. 1–7). IEEE.
70. Yang, J., Wang, Z. and Zhang, X., 2015. An ibeacon-based indoor positioning systems for hospitals. *International Journal of Smart Home*, 9(7), pp.161–168.
71. Alsinglawi, B., Elkhodr, M., Nguyen, Q.V., Gunawardana, U., Maeder, A. and Simoff, S., 2017. RFID localisation for Internet of Things smart homes: A survey. arXiv preprint arXiv:1702.02311.
72. Wang, M., Zhang, G., Zhang, C., Zhang, J. and Li, C., 2013, June. An IoT-based appliance control system for smart homes. In *2013 Fourth International Conference on Intelligent Control and Information Processing (ICICIP)* (pp. 744–747). IEEE.
73. Al-Ali, A.R., Zualkernan, I.A., Rashid, M., Gupta, R. and Alikarar, M., 2017. A smart home energy management system using IoT and big data analytics approach. *IEEE Transactions on Consumer Electronics*, 63(4), pp.426–434.
74. Rajalakshmi, A. and Shahnasser, H., 2017, September. Internet of Things using Node-Red and alexa. In *2017 17th International Symposium on Communications and Information Technologies (ISCIT)* (pp. 1–4). IEEE.
75. Alam, M., Sathaye, J. and Barnes, D., 1998. Urban household energy use in India: Efficiency and policy implications. *Energy Policy*, 26(11), pp.885–891.
76. Theodoridis, E., Mylonas, G. and Chatzigiannakis, I., 2013, July. Developing an IoT smart city framework. In *IISA 2013* (pp. 1–6). IEEE.
77. Zantalis, F., Koulouras, G., Karabetsos, S. and Kandris, D., 2019. A review of machine learning and IoT in smart transportation. *Future Internet*, 11(4), p.94.

78. Davcev, D., Mitreski, K., Trajkovic, S., Nikolovski, V. and Koteli, N., 2018, June. IoT agriculture system based on LoRaWAN. In *2018 14th IEEE International Workshop on Factory Communication Systems (WFCS)* (pp. 1–4). IEEE.
79. Strohbach, M., Ziekow, H., Gazis, V. and Akiva, N., 2015. Towards a big data analytics framework for IoT and smart city applications. In *Modeling and Processing for Next-Generation Big-Data Technologies* (pp. 257–282). Springer, Cham.
80. Khanna, A. and Anand, R., 2016, January. IoT based smart parking system. In 2016 *International Conference on Internet of Things and Applications (IOTA)* (pp. 266–270). IEEE.
81. Yun, M. and Yuxin, B., 2010, June. Research on the architecture and key technology of Internet of Things (IoT) applied on smart grid. In *2010 International Conference on Advances in Energy Engineering* (pp. 69–72). IEEE.
82. Ferdowsi, A., Challita, U. and Saad, W., 2017. Deep learning for reliable mobile edge analytics in intelligent transportation systems. arXiv preprint arXiv:1712.04135.
83. Bojarski, M., Del Testa, D., Dworakowski, D., Firner, B., Flepp, B., Goyal, P., Jackel, L.D., Monfort, M., Muller, U., Zhang, J. and Zhang, X., 2016. End to end learning for self-driving cars. arXiv preprint arXiv:1604.07316.
84. Tyagi, S., Agarwal, A. and Maheshwari, P., 2016, January. A conceptual framework for IoT-based healthcare system using cloud computing. In *2016 6th International Conference-Cloud System and Big Data Engineering (Confluence)* (pp. 503–507). IEEE.
85. Park, N. and Kang, N., 2016. Mutual authentication scheme in secure Internet of Things technology for comfortable lifestyle. *Sensors*, 16(1), p.20.
86. Kato, S., Ando, M., Kondo, T., Yoshida, Y., Honda, H. and Maruyama, S., 2018. Lifestyle intervention using Internet of Things (IoT) for the elderly: A study protocol for a randomized control trial (the BEST-LIFE study). *Nagoya Journal of Medical Science*, 80(2), p.175.
87. Dlodlo, N. and Kalezhi, J., 2015, May. The Internet of Things in agriculture for sustainable rural development. In *2015 International Conference on Emerging Trends in Networks and Computer Communications (ETNCC)* (pp. 13–18). IEEE.
88. Tzounis, A., Katsoulas, N., Bartzanas, T. and Kittas, C., 2017. Internet of Things in agriculture, recent advances and future challenges. *Biosystems Engineering*, 164, pp.31–48.
89. Meunkaewjinda, A., Kumsawat, P., Attakitmongcol, K. and Srikaew, A., 2008, May. Grape leaf disease detection from color imagery using hybrid intelligent system. In *2008 5th International Conference on Electrical Engineering/Electronics, Computer, Telecommunications and Information Technology* (Vol. 1, pp. 513–516). IEEE.
90. Sarangdhar, A.A. and Pawar, V.R., 2017, April. Machine learning regression technique for cotton leaf disease detection and controlling using IoT. In *2017 International Conference of Electronics, Communication and Aerospace Technology (ICECA)* (Vol. 2, pp. 449–454). IEEE.
91. Nandyala, C.S. and Kim, H.K., 2016. Green IoT agriculture and healthcare application (GAHA). *International Journal of Smart Home*, 10(4), pp.289–300.
92. McRae, L., Ellis, K. and Kent, M., 2018. Internet of Things (IoT): Education and technology. Relationship between education technology students with disabilities Leanne, Research (pp. 1–37).
93. Freina, L. and Ott, M., 2015, April. A literature review on immersive virtual reality in education: State of the art and perspectives. In *The International Scientific Conference elearning and Software for Education* (Vol. 1, No. 133, pp. 10–1007).
94. Lee, K., 2012. Augmented reality in education and training. *TechTrends*, 56(2), pp.13–21.
95. Liu, Y. and Wu, L., 2016. Geological disaster recognition on optical remote sensing images using deep learning. *Procedia Computer Science*, 91, pp.566–575.

96. Maeda, H., Sekimoto, Y. and Seto, T., 2016. Lightweight road manager: Smartphone-based automatic determination of road damage status by deep neural network. In *Proceedings of the 5th ACM SIGSPATIAL International Workshop on Mobile Geographic Information Systems* (pp. 37–45). ACM.

97. Advani, S., Zientara, P., Shukla, N., Okafor, I., Irick, K., Sampson, J., Datta, S. and Narayanan, V., 2017. A multitask grocery assist system for the visually impaired: Smart glasses, gloves, and shopping carts provide auditory and tactile feedback. *IEEE Consumer Electronics Magazine*, 6(1), pp.73–81.

98. Singh, B., Marks, T.K., Jones, M., Tuzel, O. and Shao, M., 2016. A multistream bi-directional recurrent neural network for fine-grained action detection. In *Proceedings of the IEEE Conference on Computer Vision and Pattern Recognition* (pp. 1961–1970).

99. Kim, H.-Y. and Kim, J.-M., 2017. A load balancing scheme based on deeplearning in IoT. *Cluster Computing*, 20(1), pp.873–878.

100. Ba, J. and Caruana, R., 2014. Do deep nets really need to be deep? In *Advances in Neural Information Processing Systems* (pp. 2654–2662).

101. Denil, M., Shakibi, B., Dinh, L., Ranzato, M. and de Freitas, N., 2013. Predicting parameters in deep learning. In *Advances in Neural Information Processing Systems* (pp. 2148–2156).

102. Lane, N.D., Bhattacharya, S., Georgiev, P., Forlivesi, C. and Kawsar, F., 2015. An early resource characterization of deep learning on wearables, smartphones and internet-of-things devices. In *Proceedings of the 2015 International Workshop on Internet of Things towards Applications* (pp. 7–12). ACM.

103. Han, S., Liu, X., Mao, H., Pu, J., Pedram, A., Horowitz, M.A. and Dally, W.J., 2016. Eie: Efficient inference engine on compressed deep neural network. arXiv preprint arXiv:1602.01528v2 [cs.CV].

104. Courbariaux, M. and Bengio, Y., 2016. Binarized neural networks: Training deep neural networks with weights and activations constrained to +1 or −1. arXiv preprint arXiv:1602.02830v3 [cs.LG].

105. Venkataramani, S., Ranjan, A., Roy, K. and Raghunathan, A., 2014. AxNN: Energy-efficient neuromorphic systems using approximate computing. In *Proceedings of the 2014 International Symposium on Low Power Electronics and Design* (pp. 27–32). ACM.

106. Moons, B., De Brabandere, B., Van Gool, L. and Verhelst, M., 2016. Energyefficient convnets through approximate computing. In *Applications of Computer Vision (WACV), 2016 IEEE Winter Conference* (pp. 1–8). IEEE.

107. Chen, Y.-H., Krishna, T., Emer, J.S. and Sze, V., 2017. Eyeriss: An energyefficient reconfigurable accelerator for deep convolutional neural networks. *IEEE Journal of Solid-State Circuits*, 52(1), pp.127–138.

108. Chen, T., Du, Z., Sun, N., Wang, J., Wu, C., Chen, Y. and Temam, O., 2014. Diannao: A small-footprint high-throughput accelerator for ubiquitous machine-learning. In *ACM Sigplan Notices* (Vol. 49, No. 4, pp. 269–284). ACM.

2 Digital Twin
Towards Internet of Drones

J. Arora and M. Tushir
Maharaja Surajmal Institute of Technology

CONTENTS

2.1 INTRODUCTION

In this modern era, where one device is controlled by the other device with its geographical impression via high-speed internet connectivity has led to development of the concept known as Internet of Things (IoT) [1]. The IoT forms an intelligent environment with the help of invisible link through the wireless network that can be monitored and programmed. The vast growth of Internet of things is visible with the fact that it is going to reach 75 billion devices, that plays a key role in fields like forest planting, environment monitoring [2], agriculture [3], and many others by 2025.

Internet of Drones (IoDs) are also one of the major applications of IoT, where the widely used Drones or Unmanned Aerial vehicles (UAVs) are joined with the Internet to better facilitate the opportunities in different fields. Gharibi et al. [4] proposed IoDs as a control architecture of layered network which help to control the UAVs. IoDs have been gaining attention for their ability to perform various tasks like disaster management, delivery of goods, real time monitoring and other industrial applications as shown in Figure 2.1.

The purpose for their success is their low maintenance cost, high maneuverability and high mobility even in remote locations and unmanned navigations. The possibility of their commercial use in broad range of applications is extensively explored by

DOI: 10.1201/9781003132868-2

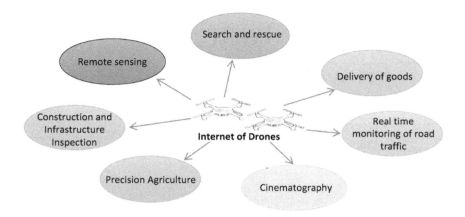

FIGURE 2.1   Applications of Internet of Drones.

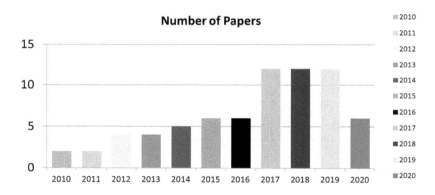

FIGURE 2.2   Distribution trends of research.

the researchers [5]. The applications of IoDs are generally operated autonomously without physical presence of the pilot, so there is a need of techniques that enable UAVs for such autonomous flights with more reliability and security.

IoDs have limitations such as UAVs maneuvering in environments that are in the presence of other mobile vehicles, robots, workers, and heavy-duty tools. It becomes a challenging task to avoid collisions or perform navigation and many other tasks. It results into the need of better positioning optimization algorithms, controlling the issues that are related to ego-motions [6].

Figure 2.2 shows the sample distribution trends of research used for this study. The application of drones increases in the research community by integrating the drones with the technological aspects of Digital Twinning. The concept of Digital Twinning helps in collecting and processing real time data by controlling virtually the physical device, using different hybrid methods which allow connecting the physical device with virtual mode of operation. The digital twin collects the on-board sensor data from the UAV and processes that information using the virtual model to create real-time predictions of the health of the vehicle. These techniques can be used to resolve the issues related to smooth functioning of IoDs. Different frameworks have been proposed to tune the parameter of the controller with the help of machine learning (ML) algorithms to better adapt to the new environment [7–12]. Various adaptive ML- and deep learning (DL)-based control algorithms have been proposed by the researchers implementing the concept of digital twinning for better control of autonomous flight in UAVs [13–17]. Researchers are introducing 'digital twins' that integrate computational models and ML to predict functioning of the UAVs and planning the path in real-time scenario for autonomous flight [18–20]. Kumar et al. [86] have proposed a neuro-fuzzy-based system to assist reliable and efficient route selection framework for IoDs. A routing framework provides better quality of services. Kurdi et al. [87] proposed a system for positioning and navigation of IoDs by implementing artificial neural network. A model is developed for planning the motion of the robot using competitive learning. Zhang et al. [88] have used different intelligent algorithms for the task of adaptive flight gesture adjustment, to avoid collision and optimization of trajectory for the process of collection of data.

The main use of ML- and DL-based techniques of IoDs in implementing 'Digital Twinning' lies in its capability to identify complex patterns from the raw input data captured by the sensors integrated in the UAV, and learning proper hierarchical representations of the underlying information at different levels. These computational models have been provided as boosters towards solving the problem related to the functioning of IoDs.

The contribution of the chapter can be summarized as follows:

1. Basic Learning Methodology of ML- and DL-based algorithms.
2. Effective use of the ML and DL approaches in the design of communication protocols and the architecture of IoT-based UAV is introduced.
3. Use of ML- and DL-based algorithms in the operations, assisted by the IoDs are overviewed.
4. Finally, some of the gaps and the challenges have been identified in the IoDs that give some open research issues for the perspective researchers.

2.2   LEARNING METHODOLOGY IN MACHINE LEARNING AND DEEP LEARNING TECHNIQUES

ML is a concept of Artificial Intelligence that enhances system capability to learn automatically from the knowledge hidden within the information available in the data [21]. It focuses on the development of programs that can learn from the data without being explicitly programmed. The nature of this learning mechanism depends on the availability of the data or observations, direct experience or instructions, to discover pattern in the data. ML algorithms are categorized as follows:

i. **Supervised Learning (SL):** In this learning mechanism, a model is trained with the help of labeled data to predict future events from the learning experience. Here the model generates an inferred function to make future predictions. The model is used to provide outputs for any new input after adequate training. The performance is measured on the basis of accuracy verified from the intended output, errors are calculated and the model is retrained in order to improve the results. They include method of classification and regression. CNN-based DL methods were used for the extraction of useful information from the data collected by different sensors [89].

ii. **Unsupervised Learning (UL):** The information provided in these algorithms is neither labeled nor it can be classified. In this learning mechanism, a mathematical model is applied to infer hidden patterns from the unlabeled data. The algorithms explore the data to infer some information present in the datasets. They include methods such as clustering and dimensionality reduction. The concept of UL was used for the exposure of active spying in the relay-based network of UAV and was detected using SVM and k-means clustering. Here the authentication was carried by uplink [92]. The training datasets are used to facilitate the process of training, using the knowledge of statistics and wireless signals. The experiment results proved the performance of one-class SVM as more stable as compared to k-means.

iii. **Semi-Supervised Learning (SSL):** In semi-supervised learning, the training data involves a small percentage of labeled data and a large percentage of unlabeled data. This learning is chosen when the acquired labeled data requires some skilled resources to provide a good estimate to parameters of unsupervised learning process. UL was used for the purpose of object recognition [90]. It has some requirement of labeled data. It not only requires human intervention, but it is also difficult to process.

iv. **Reinforcement Learning (RL):** In this learning mechanism, a model interacts with its surroundings by generating events and determining errors or rewards. Different trial and error search methods and feedback methods form the features of reinforcement learning. This method permits machines to automatically discover the ideal behavior within a precise scenario in a way to maximize performance. In RL, reinforcement signal is generated on the basis of reward feedback mechanism. They use the method of value iteration and policy iteration. RL comprises different processes with optimized outcomes during the whole stated space in a way to attain maximum overall reward. Problems related to robotics are often based on such tasks with sequential structure. The problems involving such tasks are generally allowed to be solved by using reinforcement learning framework [91]. Ma et al. [96] proposed an algorithm for the purpose of obstacle avoidance by the use of RL.

With the advancement in the field of ML, it gave rise to another sub-field of ML algorithms, known as DL algorithms. DL refers to the process of learning representations of input data primarily through the transformations. This makes it easy to perform classification to prediction tasks. Figure 2.3 gives the basic difference between ML

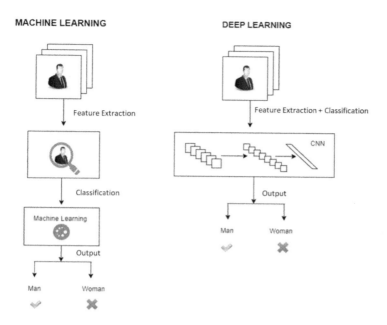

FIGURE 2.3   Machine Learning Vs Deep Learning.

and DL algorithms. ML algorithms require humans for the task of feature extraction from the data, while DL algorithms do not require human interventions as multi-level layers of convolution neural networks (CNN) perform the task of feature extraction and classification [97–99]. There are a number of ML and DL algorithms that are proving successful in solving the complex task by analyzing the data collected from the sensors of the drones and help to perform complex tasks with better accuracy. There are several popular DL networks that include- Recursive Neural Networks (RvNN), Recurrent Neural Networks (RNN), Convolutional Neural Networks (CNN), and Deep Generative Networks (DBN, DBM, GAN, VAE). Table 2.1 summarizes the different DL networks.

2.3   COMMUNICATION ARCHITECTURE OF INTERNET OF DRONES

This section provides a detailed layered architecture of Internet of Drones (IoDs). Layering helps to inculcate many features such as scalability, ease of maintenance for modifying the layer, separation of concerns, and more flexibility with the independence to other layers. The architecture provides guidance for the functioning of the UAVs [29] and how these layers have influenced by the performance of different ML- and DL-based algorithms. The architecture presented in Figure 2.4 helps to better understand the functioning of the UAV with respect to different components participating in the working of the aerial vehicle and how the communication is established at global and human interface level [30].

TABLE 2.1

Summary of Various Deep Learning Algorithms

Deep Learning Network	• Summary	Advantages	Limitations	Application Area
Recursive Neural Networks [22]	In this type of deep neural network, same set of weights are applied recursively over a structured input, particularly directed acyclic graphs, to give an output.	Uses a tree like structure and is a generalized Recurrent Neural Network.	There is difficulty in training the network.	Preferred DL network for NLP.
Recurrent Neural Networks [23]	This DL network consists of processing in which output from previous steps is set as input to current step. It contains feedback loops.	It is useful in time series prediction as it remembers all the information through time.	RNN faces gradient vanishing and exploding problems. There is difficulty in training the network.	Preferred DL network for NLP and speech processing.
Convolutional Neural Networks [24]	These networks are used in the image recognition task, specifically designed for pixel data.	These can achieve high accuracy in image recognition task.	Using CNN on data with no local coherence would produce poor results. Large quantity of training data is required. It is a blackbox technique	Preferred for image recognition, NLP, computer vision, speech recognition, etc.
Deep Generative Networks	Deep generative models are essentially of four types (DBN [25], DBM [26], GAN [27], VAE [28]). Generative models define procedures that basically are used in production of samples of data.	These are capable of learning representations, handling large amounts of unlabeled data and handling exploitation/exploration tradeoffs.		These are applied in computer vision, NLP, graph mining, and reinforcement learning.

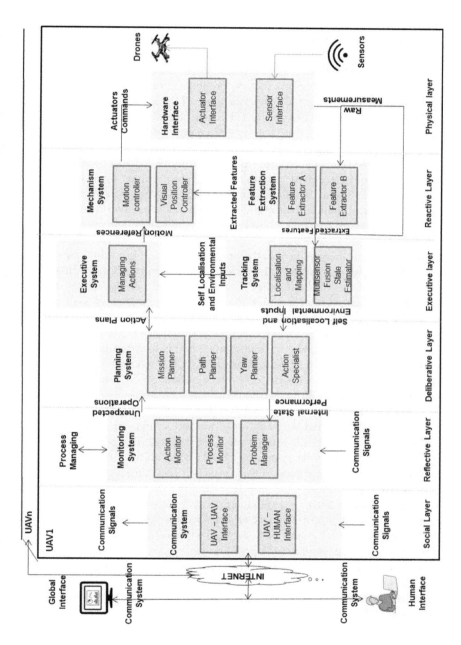

FIGURE 2.4 Communication architecture of Internet of Drones.

The architecture comprises of n number of unmanned aerial vehicles that can be categorized into different interfaces and modules.

i. **Hardware interfaces**: This module includes hardware components of system like sensors, actuators and other important components. There are different primary and secondary components required by UAV to operate. Primary component includes batteries, camera, GPS module, and flight controller. Secondary components include propellers, solar panel, and transceivers. These components allow performing all the actions and reactions. For the complex implementation of the DL algorithms, advancement in chip fabrication technology enabled solving design problems at an effective cost and within a time frame that was unthinkable [31–34] but was required.

ii. **Mechanism system**: This system deals with motion control subsystem which specifically receives commands of desired values for a variable (position, orientation, or speed) from different types of sensors and special cameras installed through extracted features interface. These required values are translated into low-level commands that were sent to the actuators.

iii. **Feature extraction system**: It is the process of extracting useful and reliable information or features of the data obtained from sensors like image, temperature, pressure or any other. The major job of DL algorithms is to learn from structural representation of data, so feature extraction system is somewhat controlled by the DL algorithms. Vincent et al. proposed Stacked Denoising Autoencoders (SDAE) as a feature extractor. Here denoising autoencoders were trained with higher noise levels to strain the model to mine more distinct and less local features [35]. CNN features were applied on the captured images for better object recognition [36–40] and classification of the scenes [41–43].

iv. **Tracking system**: This system involves reception or generation of sensor's data to understand surrounding information and get environment knowledge or creating a localized map. An example component within the situational awareness system is SLAM algorithms. Bry et al. [44] used SLAM for obstacle recognition in unfamiliar quarters, using scanning lidars.

v. **Executive system**: This system manages action by various other sections and the action by itself. It deals with overall action flow or sequence management of the system. Xiang et al. [45] for UAV-based tracking used robust consensus-based algorithm for object tracking. He used neural network-based ML approach to improve performance of target recognition.

vi. **Planning system**: This process provides solution to the task that has been provided to the system by involving information from tracking and monitoring interfaces. Examples like solution planning or path planning are done using different ML-based approaches for optimal path planning.

vii. **Monitoring system**: This system ensures development done by drone/robot in positive direction that is toward the goal. In monitoring system, a self-supervision process is considered where self-assessment of task provided is monitored by the system itself, whether it will be dealing with errors, fault lines or any sort of recovery options.

viii. **Communication system**: This system develops interaction between an operator and drone or signaling between base station and UAV for decision making purpose. In the IoDs, the high-speed internet is used to develop a communication link between UAV and the operator. To develop the interface between human-UAV and UAV-UAV, various types of protocols are used such as Wi-Fi, Wi-Max, Long-Term Evolution-Advanced (LTE-A), 5G, etc. Table 2.2 gives the details of different communication protocols and technologies used for IoT-based UAV.

These communication technologies proved to perform better when applied with ML-based approaches. The integration of ML algorithms with the Wi-Fi-operated IoDs helped mobile users to access wireless, reliable and measurable operations. ML-based approaches, such as ANN is used to forecast the performance of WiMAX

TABLE 2.2
Communication Protocols and Technology

Technology	Protocols	Advantages
Wi-Fi [46,47] Up to 100 mt	IEEE 802.11a/b/g/n/ac	• Allow People to utilize their devices to control UAV.
LTE [48]	Global System of Mobile (GSM) and Universal Mobile Telecommunications System (UMTS)	• High speed data transfer between mobile devices.
LTE-A [49]	Advanced Version of LTE	• Lower latencies • Higher throughput • Improved coverage • Monitoring architecture for streaming real-time video surveillance.
WIMAX [50]	IEEE 802.16e	• Increase the size of service area and the number of communication channels among the service providers • Data transmission • VoIP services.
ZigBee [51] Up to 50 mt	IEEE 802.15.4	• Low power consumption • Low data rate • Efficient for short-range communication.
LoRa WAN [52]		• Low power consumption • Low bit rate • Long-range wireless technology.
5G [53,95]		• Low latencies • High data rates • Energy efficiency • Reduce the chance of collision and actions.

by testing on maximum and minimum number of users [50]. ZigBee [51] in combination with the ML techniques proved to provide better performance in the area with short-range communications such as smart farming [54] and libraries [55]. ANN integrated with LoRaWAN helps to find the cost of propagation [52], as they are less demanding computationally than deterministic and non-deterministic techniques. 5G networks have proved to provide one of the most cost effective and energy efficient solutions to the IoT-based flying objects [56,95]. Moreover, integrating DL-based solutions to the IoDs platforms provided more secure transmission over the 5G networks [57].

Further, in this communication architecture a UAV is connected to n-number of other UAVs. UAV is communicating with the UAV and UAV is communicating with human communication system. These connections can be made in the form of centralized or decentralized ways. In the centralized architecture, all UAVs are connected to the base station. In decentralized architecture, all UAVs are connected to each other, directly or indirectly without the help of base station as the central node.

2.4 CONTRIBUTION OF MACHINE LEARNING AND DEEP LEARNING ALGORITHMS TOWARD INTERNET OF DRONE OPERATIONS

IoDs lead to the concept of autonomous approach of driving. It uses the concept of flying ad hoc networks as the ability to be self-guided, without the involvement of human. They are deployed on open wireless sensor networks with an increased requirement of real time data processing and intelligent tools to handle the complexity of rapidly changing topology in mobile environment. There are different operations performed by UAVs, such as surveillance, object detection, collision avoidance, navigation, data collection, tracking, and environmental monitoring, among others. To employ the IoDs with such features, ML and DL algorithms have contributed toward a smooth functioning of various operations related to IoDs. Figure 2.5 depicts the operations discussed in the paper that are enhanced by the support of ML and DL algorithms. The control strategies are divided into four categories: navigation, battery scheduling, object detection, and collision avoidance and security.

2.4.1 NAVIGATION

With the shift of technology towards more real-world problems and advanced flying objects, the need for real time path planning becomes a fundamental and basic requirement for UAVs. The application of IoDs for civil operations such as foresting, health care, surveillance and rescue operations, UAVs are required to opt for more optimal and safe paths. UAVs with optimal navigation abilities are critical to implement. The main aim of navigation is to plan a path from source to destination, avoiding all the intermediate obstructions. Recently, researchers have adapted different ML approaches in the field to provide more optimal solutions. Junell et al. [58] proposed an approach to calculate optimal path planning, using the concept of reinforcement learning (RL). RL can be considered as an effective approach where the

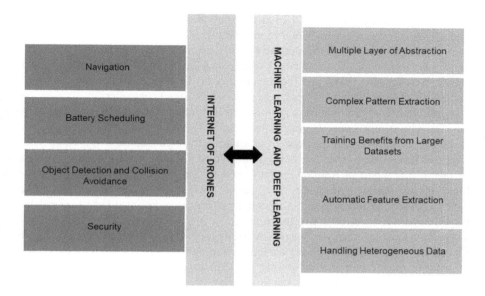

FIGURE 2.5 Machine Learning and Deep Learning towards Internet of Drones.

decision has to be taken in the unknown environment, based on the feedback method. However, this work was limited to discrete and unchangeable surroundings. Zhang et al. [59] have given the concept of cooperative and geometric learning algorithm (CGLA), designed for planning of path for multiple UAVs sharing the information. In CGLA, parameters were tuned to optimize the cost of path and to avoid collision between the UAVs flying in the multi-agent mode. Every time cost matrix and weight matrix are updated to calculate the optimal results.

Kan et al. [19] defined a navigation approach for terrain-based UAVs by utilizing the concept of extreme learning machine (ELM) algorithm. This approach allows capturing terrain-based information from the sensors present on board for the purpose of planning navigation when global positioning system fails due to environmental constraints. ELM algorithm is optimal and efficient as it uses single layer feed forward network and helps to decrease training time of Neural Networks. The experimental results of ELM algorithm proved to perform efficiently with less time of computation for computing multi-resolution terrain and terrain with different sizes.

Aznar et al. [60] used CNN approach to learn from the images to perform position-dependent actions. A potential function was developed to plan a global position map. The process involves registration of the image followed by generation of control action for planning the movement to be carried out. This behavior is encoded using CNN, and has proved to give better results as compared to other image registration techniques.

Giusti et al. [77] proposed a DNN-based approach for navigation of UAVs in unstructured environment. The cluttered natural scenario of dense forest is considered. A model was trained with DNN training method to map image to perform different actions (such as turn left, go straight, or turn right), with a final SoftMax layer.

In 2018, Loquercio et al. [20] tried to improve the working of civilian drones. The model was trained using CNN to navigate through urban roads and called this approach DroNets. Though this approach didn't require a real-world map of environment but consumed a lot of data which was acquired from other vehicles like car and bicycle to safeguard UAVs and the dynamicity of the path planning was not fully implemented or utilized as compared to other CNN-based controllers.

Different ML algorithms were proposed or integrated with existing techniques to tune the parameters related to generating control.

Jardine et al. [61] used the concept of learning through automata that make multiple interactions with unknown or random environment to choose efficient action to be taken; it is a part of Reinforcement Learning. Taking in consideration the nonlinear noises, various parameters of controller were changed for a nonlinear UAV, which not only helped in position tracking performance, but also led to optimized weighing parameter of objective function.

Neural Network (NN) is based on exemplary behavior of the human brain functioning that has always been an interesting part of Artificial Intelligence. So, in order to mimic human functioning in UAVs for adaptive control while taking autonomous flights, researchers try to handle uncertainty and non-linear dynamics of system using NN. NN contains many connected processors called neurons which are modelled as weights. They help in complex data representation and processing. NN are widely used in solving complex problems. In this field too, multiple researchers like Kurnaz et al. [13] used them in FNN to develop an adaptive neuro-fuzzy inference system (ANFIS)-based controller. Lin et al. [15] also used recurrent NN for intelligent control system that can have dynamic control and information storing ability.

Chowdhary and Johnson introduced concurrent learning [14] adaptive controller that was targeted at improving the convergence of parameter in model reference adaptive control (MRAC). Later,, Chowdhary et al. [62] went on to utilize concurrent learning to control transfer problem that occurs between systems having alike control structure. He employed two learning methods based on MRAC laws: (i) concurrent learning MRAC and (ii) budget Kernel restructuring concurrent learning.

Bansal et al. [63] aimed at creating an NN-based dynamic model that can be used to learn the dynamics of UAV for tracking those trajectories over which the NN model was not trained using NN. Later on, in experiments it was suggested that even simple NN as FNN was also able to learn dynamics of UAV with good accuracy. Hence, allowing future improvements where generalized capabilities of NN could be used with pre-determined knowledge about the system came into the picture.

As seen from the above work, a lot of work toward improving navigation of UAVs is carried out using different ML and DL approaches. But still a lot of work needs to be carried out for improving the problem of localization and proper path planning.

2.4.2 BATTERY SCHEDULING

In Scope of Drones, battery issues that prevent free exploration of a drone are discussed. Stationary charging further does not allow the usage of drones in specific harsh conditions or for long-term durations, whereas mobile charging points provide a solution but has its own set of problems like scheduling between multiple drones from a mobile charging station.

In Ref. [64], the authors focus on solving drone scheduling as integer scheduling programming problem and battery assignment by heuristic approach, given that the battery is swiftly shifted. Various drones share their information with central charging point among a distributed network to function, though in the proposed method, the necessity of a stationary service station acts as a limitation. The authors proposed auction method which was further taken forward in Ref. [65] with two auction mechanisms to get optimal solutions within distributed scheduling, although they required prior knowledge of environment which limits its application. Shin et al. [66] proposed a DL solution to overcome the limits of prior knowledge and scheduling problem, by using auction mechanism. Here, time slots for charging the batteries are auctioned by the process of bidding. The DL model enables the knowledge about distribution of drones, participating in the process. Here, the charging station acts as the leader of the distribution mechanism and coordinates with other vehicles through Internet of Vehicles. The proposed model uses ReLU as the activation function and SoftMax as the classification function. The proposed approach proved to be less complex and require minimum computation time.

Li et al. [84] proposed on-board deep-Q-network for online transfer of power and collection of data. They proposed to develop a process to minimize the overall loss of data packet, used for sensing of devices, by optimally determining the device to be charged and interrogated for data collection, and the instantaneous patrolling velocity of the UAV. A Markov Decision Process (MDP) works with states of battery level and data queue length of sensing devices, channel conditions, and waypoints given the trajectory of the UAV, and solves it optimally with the help of Q-learning.

Liu et al. [85] proposed an ML-based method for designing of trajectory and control of power for multiple UAV-assisted wireless networks. The Q-learning-based method is used to determine optimal position of the UAVs, and Echo state network-based prediction algorithm is used to predict future positions of the UAVs. Lastly, a multi-agent Q-learning-based algorithm is used for estimating the location of UAVs periodically based on the movement of users.

2.4.3 Object Detection and Collision Avoidance

The prerequisite for UAVs is to be fully autonomous and perform operation, such as ability to carry out emergency landing, avoid collision, and be equipped with sense beyond line-of-sight operation. Locating path and authenticating communication are very important aspects of IoDs system navigation. Similarly, an unmanned vehicle needs to understand other real-world challenges like object detection and collision avoidance. There are two major approaches, namely, co-operative and non-co-operative approaches that define the communication between existing UAVs in the environment. In cooperative drone [67] positioning measurement method, sensor data fusion is used via IoDs to obtain optimized positioning information. The principle of sensor data fusion is to obtain measuring information from sensors and then analyzing the information, using ML and deep-learning approaches to have better localization. In non-cooperative approach, information about the object is detected using acquired training dataset.

Bachrach et al. [68] used a Microsoft Kinetic camera in an unknown room, and Zhang et al. [69] tried object tracking by 3D localization space using DL approach. The system uses a method for multi-object tracking known as TrackletNet Tracker (TNT) which considers temporal and pictorial information to track the objects positioned on the ground for UAV applications. They also used multi-view stereo technique to localize the objects which are tracked. Fraundorfer et al. [70] proposed using stereo camera for efficient depth estimation and localization. These algorithms were early steps in the field and required much work for improving computing efficiency for UAVs, that's when ML algorithm were utilized in the zone. Rabah et al. [71] proposed a method for the purpose of object tracking and detection. Here, the UAV control software is executed on a heterogeneous resource allowing limiting the use of computing units on a UAV. A convolutional neural network (CNN) is used for the purpose of object detection. For the purpose of object tracking, an algorithm is designed for tracking of object via stream of sequential. Further tracking unit and dynamics of UAV are enhanced by the use of gain-scheduled PID controller. AlGuo et al. [72] used support vector machine and Gaussian mixture model to sense landing spot where they worked on an automatic visual image-based aerial image classification and compared with the baseline. Reinforcement learning algorithm was also used by Junell et al. [58], where they repeatedly trained the model to learn an efficient, converged value function in order to take picture of disaster site.

ML approaches require to extract features from the training data in order to specify the descriptor to perform the task of learning using basic classification model or regression models. In comparison to ML methods, DL methods provide solution to this, as using CNN, the features are extracted automatically by feeding the raw data into the network. DL algorithms proved to provide better solution while extracting features from the data acquired by the sensors in the real environment [73]. The DL proved to provide an effective autonomous collision avoidance technique, in the large number of robotic operations, such as object detection, collision avoidance, navigation and control by the use of complex learning approaches.

There are a number of DL techniques used for the purpose of collision avoidance which use the information captured by the cameras present in the UAVs and generate a possible set of actions to be carried out [74,75]. These approaches are based on reinforcement learning as they are trained on the input of expert pilots at a given situation in real world. Further, some actions are carried out by a number of intermediates' representation, estimating position of the final obstacle. Further, DL-based models are used for generating feature maps in the module for situational awareness, related to the state of robotic system and its surroundings. Such computed feature maps help to take the final decision-making process.

Researchers wanted to calculate the probability of collision of UAV with objects, including estimated uncertainty due to high speed. They reached the conclusion that their reinforcement learning algorithm without uncertainty had more collision than the one with uncertainty. Giusti et al. [77] solved this problem of pattern recognition through image classification, using DNN and trained model from real-world dataset collected from human hikers. They achieved 95% of accuracy, yet had few setbacks in perceiving trails when trails' centerline lacked space. Barták et al. [78] used four approaches, namely, hierarchical clustering, K-means clustering, Gaussian mixture model, and hidden Markov models, though their approach lacked accuracy in diverse environments and couldn't communicate to multi-command.

Wang et al. [94] proposed a UAV-based system for tracking and recognizing a target with the help of an intelligent gimbal system with the proficiencies of fast image processing, accurate camera positioning, and fusion of information using multi-modality process. Different algorithms are integrated for tracking of target, moving background processing, and neural network, based on detection of target and recognition of target, that resulted into optimized approach. Further, information related to geo-location, contextual and environment is embedded using a geographic information system (GIS). The experimental results proved the robustness of the proposed technique for the purpose of target tracking in UAVs.

2.4.4 SECURITY

With more and more advancement toward UAV's extensive usage in fields of military, next-generation applications such as smart parking, smart tracking, and other important fields, the concerns regarding UAV's security and threats to their safety both physically and algorithmically will keep rising. Various researchers had worked in the field to ensure that more and more secure, incorruptible UAVs could be built and implemented. Providing authentication and security is the key issue and concern to UAVs in such autonomous environment. Karimibiuki et al. [79] focused on considering authentication failure as a threat and tried to implement the feature of providing authentication using basic ML approaches such as K-Nearest Neighbor (K-NN), Support Vector Machine (SVM), and Logical Regression (LR). They consider if the attacker followed man in middle attack, where the attacker records and learns communication controls, and changes them and conveys wrong message to the UAV. The attacker mainly wants to crash or dump the UAV to prevent surveillance. So, they train their model using three different algorithms to correctly detect the right path (precision) and the wrong path (recall).

KNN stores all the flight trajectories. All flights are classified on the basis of distance calculation and K-nearest neighboring tracks. SVMs are non-probabilistic binary linear classifier supervised learning models. They apply statistics of support vector to categorize unlabeled data and can predict next point of category. Here, by finding a hyper-plane to maximize the margin between two classes, thus tracking pattern anomaly as soon as UAV goes to false territory. LR is used during analysis when dependent variable is binary in nature. It attempts to set up a relationship between characters and creates a probability of the next state.

On the basis of the experiments, the data of GCS such as altitude, latitude, longitude, and time-series information of drone's attitude such as yaw, roll, and pitch is included. They applied percentage split of 80% and discovered that KNN is the best among the three, when it comes to recall and slightly behind SVM in precision metric. At the same time, linear regression is the worst overall.

DL and ML algorithms can provide a great advantage in the classification of various security threats and events. Das and Ghosh [80] proposed a DL framework for reconstructing the missing data in remote sensing analysis. Many authors [81] studied various methods for detection and identification of drones using emerging DL methods. They proposed techniques for exploiting unique acoustic fingerprints of UAVs to classify them by feature detection. But still, to secure the UAVs from attacks and authentication issues remains a key challenge. Ye et al. [82] have defined a method for secure transmission of information between multiple UAVs in the network by calculating secrecy outage probability and average secrecy capacity using Monte Carlo simulations. Challita et al. [57] emphasized on different problems related to the wireless connectivity of the IoDs and the security issues involved. To overcome these challenges, different solutions are proposed on the basis of approaches using ANN models. This method proved to provide more secure transmission over 5G networks.

Challita et al. [83] proposed a path planning technique for a network of cellular connected UAVs to control interference among the UAVs. In this technique, deep reinforcement learning is used based on Echo State Network (ESN) of cells. The proposed deep ESN framework is trained to permit each UAV to plan every observation of the network state to a response, with the aim of optimizing a sequence of time-dependent utility functions. Each UAV in the network uses ESN to calculate its optimal path, but the power is required for transmission and different associations with the cell.

Hoang et al. [92] proposed a UL approach for authentication of UAV-based relay network for detection of active eavesdropping. Yue et al. [93] proposed the distributed system for locating and targeting the security and privacy issues of the UAV controlled by amateurs over some sensitive areas. The system uses wireless acoustic sensors along with SVM-based ML algorithms. Then the proposed system was analyzed over telemetry data of such UAVs. They used a software-defined radio (SDR) platform implementing an SVM-based ML approach, originating through classifiers to identify if a packet transmitted came from a UAV. On the results of analysis, malicious behaviors of UAVs are detected and controlled by employing other surveillance UAVs.

2.5   RESEARCH CHALLENGES AND FUTURE INSIGHTS

ML and DL algorithms have provided a promising support to various operations of controlling autonomously functioning UAVs due to their various features involving learning from raw data by extracting complex pattern in a self-learning environment. DL algorithms require a large set of training data for extracting useful information from the images. Although they provide a promising solution, there are a number of challenges required to be focused on—as a new research area for advancement in the working of IoDs.

- **Data integration and standardization**: With the use of Deep Learning approaches, a system has to deal with big data. In the multiple UAVs flying mode, the data is to be collected from numerous sensors simultaneously. The data transmission, processing, and information gathering in the aerial network is a challenge to be resolved in the multi-cloud environment. Further, as data is to be collected from multiple resources, there are no generalized standards in terms of communication between the drones as well as platforms and algorithms. Efficient algorithms should be suggested for effective utilization of the intelligent systems, and big data processing with low computation.

- **Security and privacy**: Maintaining security of the data collected at the global level is one of the key concerns. Due to the lack of the on-board encryption mechanism, the data is very easily accessed by the hackers. A safeguard mechanism is required to be built using various encryption techniques, optimization techniques for image acquisition and processing techniques such as stitching of aerial image, and fusion of data need to be implemented. The drones are widely used by criminals and terrorists in order to breach different security features. They should be integrated with multi-dimensional authentication and authorization features.

- **Hardware implementation challenge**: The implementation of DL and ML algorithms has been a hot topic in research. These algorithms require a lot of processing and high-level fabrication technology which enabled solving on-board design problem for effective utilization. Highly integrated chips are required at a low cost to allow these technologies to be effectively used.

- **Ego-motion estimation**: Independent movement of one or more aerial vehicles in extreme situations gives rise to several problems that need to be addressed for estimation of the ego motion. To solve this issue, competently, deeply complex, but structured learning algorithms are required for analyzing the features in combination with optical flow.

- **Vehicle collision detection**: In order to avoid collision between autonomous flight vehicles and provide better safety system, approaches based on the combination of classifiers and ego-motion-based concepts should be used to promote detection rate. To avoid collision, different viewing angles can be used for learning the objects. Further, a common operating picture containing position altitude and velocity of objects around the drones should be processed.

- **Augmentation control**: The drones should be facilitated with certain additional commands when required during emergency situations or when appropriate authorization is required. Presence of good communication channel to keep in contact with the remote server and for providing better situation awareness to adapt for varying situations, context data collected from the variety of sensors needs processing using advanced learning techniques.

- **Energy drainage:** Performing operations using deep neural-network-based classification process requires a lot of processing. Here energy efficiency can be considered as performance bottleneck. So, an energy-efficient neural network should be designed to overcome the challenge of energy drainage.

2.6 CONCLUSION

The universal deployment of digital twin technologies across IoDs has developed various perceptive and autonomous capabilities toward different control strategies such as object detection, navigation, security, collision avoidance, and backup. These approaches help to deal with expansive growth of big data solutions. Basic ML algorithms have been used for security and object detection. DL algorithms are more widely applied to the feature of extraction and the object detection system of IoDs. With the collection of large data from various sensors and cameras, these algorithms help to resolve the issues of analysis of data in a more optimum way. However, in the systems with an advance level of generalizations, such as UAV monitoring, security systems need to be developed robustly by the research community. These systems implement autonomous and complex behaviors that have to be learned using supervised learning technique. The biggest challenge toward DL algorithms is the requirement of the labeled data for the process of training and learning. The efforts have been made toward learning the model using unsupervised learning methods as the environment generated a lot of unlabeled data. Making models to explore information from these datasets will generate technologically less-expensive results. Clearly, the Digital Twin-based integrated techniques are not a privilege, but rather a necessity for future services of drones, operated in wireless networks. ML-driven internet of drones will provide a path toward development of highly autonomous and secure a set of network functions and wireless services.

REFERENCES

1. A. Solanki and A. Nayyar, "Green Internet of Things (G-IoT): ICT technologies, principles, applications, projects, and challenges." *Handbook of Research on Big Data and the IoT*, IGI Global, 379–405, 2019.
2. O. Blanco-Novoa, T. M. Fernández-Caramés, P. Fraga-Lamas and L. Castedo, "A cost-effective IoT system for monitoring indoor radon gas concentration." *Sensors (Basel)*, 18(7): 2198, 2018.
3. A. Nayyar, B.-L. Nguyen and N. G. Nguyen, "The Internet of Drone Things (IoDT): Future envision of smart drones." *First International Conference on Sustainable Technologies for Computational Intelligence*, 563–580, 2020.
4. M. Gharibi, R. Boutaba and S. L. Waslander, "Internet of Drones." *IEEE Access*, 4: 1148–1162, 2016.
5. R. J. Hall. "An Internet of Drones." *IEEE Internet Compt.*, 20(3): 68–73, 2016.
6. N. H. Khan and A. Adnan, "Ego-motion estimation concepts, algorithms and challenge: An Overview." *Multimedia Tools and Appl.*, 76: 16581–16603, 2017.
7. M. Ö. Efe, "Neural network assisted computationally simple $PI^{\lambda}D^{\mu}$ control of a quadrotor UAV." *IEEE Trans IndInformat*, 7(2): 354–361, 2011.
8. S. Wang, B. Li and Q. Geng, "Research of RBF neural network PID control algorithm for longitudinal channel control of small UAV." In *10th IEEE International Conference on Control and Automation (ICCA)*, IEEE, 1824–1827, 2013.
9. W. N. Gao, J. L. Fan and Y. N. Li, "Research on neural network PID control algorithm for a quadrotor." *Appl Mech Mater*, 719(0): 346–351, 2015.
10. S. R. B. Santos, S. N. Givigi and C. L. Nascimento, "Nonlinear tracking and aggressive maneuver controllers for quad-rotor robots using learning automata." In *2012 IEEE International Systems Conference (SysCon)*, IEEE, 1–8, 2012.
11. S. R. B. Santos, S. N. Givigi and C. L. N. Júnior, "An experimental validation of reinforcement learning applied to the position control of UAVs." In *2012 IEEE International Conference on Systems, Man and Cybernetics (SMC)*, IEEE, 2796–2802, 2012.
12. P. T. Jardine, S. Givigi and S. Yousefi, "Parameter tuning for prediction-based quadcopter trajectory planning using learning automata." *IFAC-PapersOnLine*, 50(1): 2341–2346, 2017.

13. S. Kurnaz, O. Cetin and O. Kaynak, "Adaptive neuro-fuzzy inference system based autonomous flight control of unmanned air vehicles." *Expert Syst Appl*, 37(2): 1229–1234, 2010.

14. G. V. Chowdhary and E. N. Johnson, "Theory and flight-test validation of a concurrent-learning adaptive controller." *J Guid Control Dyn*, 34(2): 592–607, 2011.

15. C.-M. Lin, C.-F. Tai and C.-C. Chung, "Intelligent control system design for UAV using a recurrent wavelet neural network." *Neural Comput Appl*, 24(2): 487–496, 2014.

16. A. Punjani and P. Abbee, "Deep learning helicopter dynamics models." In *2015 IEEE International Conference on Robotics and Automation (ICRA)*, IEEE, 3223–3230, 2015.

17. J. Shin, H. J. Kim and Y. Kim, "Adaptive support vector regression for UAV flight control." *Neural Netw*, 24(1): 109–120, 2011.

18. H. Xu, Y. Gao and F. Yu, "End-to-end learning of driving models from large-scale video datasets." In *(Proceedings of) IEEE International Conference on Computer Vision and Pattern Recognition, IEEE Xplore*, 2174–2182, 2017.

19. E. M. Kan, M. H. Lim and Y. S. Ong, "Extreme learning machine terrain-based navigation for unmanned aerial vehicles." *Neural Compt. Appl.*, 22(3–4): 469–477, 2013.

20. A. Loquercio, A. I. Maqueda and C. R. DelBlanco, "Dronet: Learning to fly by driving." *IEEE Robot Automat Lett*, 3(2): 1088–1095, 2018.

21. J. Arora and M. Tushir, "A new semi-supervised intuitionistic fuzzy c-means clustering." *EAI Endorsed Transactions on Scalable Information Systems*, 1–12, 2019.

22. O. Irsoy and C. Cardie, "Deep recursive neural networks for compositionality in language." In *NIPS'14 Proceedings of the 27th International Conference on Neural Information Processing Systems*, vol. 2, 2096–2104, 2014.

23. D. E. Rumelhart, G. E. Hinton and R. J. Williams, "Learning representations by back-propagating errors." In *Nature*, vol. 323(6088), pp. 533–536, 1986.

24. Y. LeCun, B. Boser, J. S. Denker, D. Henderson, R. E. Howard, W. Hubbard and L. D. Jackel, "Back propagation applied to handwritten zipcode recognition." *Neural Compt.*, 1(4): 541–551, 1989.

25. E. H. Geoffrey, "Deep belief networks." *Scholarpedia*, 4(5): 5947, 2009.

26. R. Salakhutdinov and G. Hinton. "Deep Boltzmann machines." In *Artificial Intelligence and Statistics*, PMLR, 448–455, 2009.

27. I. Goodfellow, J. Pouget-Abadie, M. Mirza, B. Xu, D. Warde-Farley, S. Ozair, A. Courville and Y. Bengio, "Generative adversarial nets." In *Advances in Neural Information Processing Systems*, Curran Associates, 2672–2680, 2014.

28. D. P. Kingma and M. Welling. "Auto-encoding variational bayes." CoRR abs/1312.6114, 2013.

29. J. L. Sanchez-Lopez, M. Molina, H. Bavle, C. Sampedro, R. A. S. Fernandez and P. Campoy, "A multilayered component-based approach for the development of aerial robotic systems: The aerostack framework." *Journal of Int. & Robotic Syst.*, 88: 683–709, 2017.

30. A.Carrio, C. Sampedro, A. Rodriguez-Ramos, and P. Campoy, "A review of deep learning methods and applications for unmanned aerial vehicles." *Journal of Sens.*, Hindawi, 2017, Article ID 3296874, p. 13, 2017

31. B. Broecker, K. Tuyls and J. Butterworth, "Distance-based multi-robot coordination on pocket drones." In *Proceedings of 2018 IEEE International Conference on Robotics and Automation (ICRA)*, Brisbane, QLD, Australia, 6389–6394, 2018.

32. M. A. Anwar and A. Raychowdhury, " NavREn-Rl: Learning to fly in real environment via end-to-end deep reinforcement learning using monocular images." In *Proceedings of the 2018 25th International Conference on Mechatronics and Machine Vision in Practice (M2VIP)*, Stuttgart, Germany, pp. 20–22, November 2018.

33. S. Jung, S. Hwang, H. Shin and D. H. Him, "Perception, guidance, and navigation for indoor autonomous drone racing using deep learning." *IEEE Robot Autom Lett*, 3, pp. 2539–2544, 2018.

34. S. Junoh and N. Aouf, "Person classification leveraging convolutional neural network for obstacle avoidance via unmanned aerial vehicles." In *Proceedings of the 2017 Workshop on Research, Education and Development of Unmanned Aerial Systems (RED-UAS)*, Linkoping, Sweden, pp. 168–173, 3–5 October 2017.

35. R. Girshick, J. Donahue, T. Darrell and J. Malik, "Rich feature hierarchies for accurate object detection and semantic segmentation." In *Proceedings of the 27th IEEE Conference on Computer Vision and Pattern Recognition (CVPR'14)*, Columbus, Ohio, USA, pp. 580–587, June 2014.

36. R. Girshick, "Fast R-CNN," In *Proceedings of the 15th IEEE International Conference on Computer Vision (ICCV '15)*, pp. 1440–1448, December 2015.

37. S. Ren, K. He, R. Girshick and J. Sun, "Faster R-CNN: Towards real-time object detection with region proposal networks." In *Advances in Neural Information Processing Systems*, vol. 28, pp. 91–99, 2015.

38. J. Lee, J. Wang, D. Crandall, S. ˇSabanovic and G. Fox, "Realtime, cloud-based object detection for unmanned aerial vehicles." In *Proceedings of the 1st IEEE International Conference on Robotic Computing (IRC)*, Taichung, Taiwan, pp. 36–43, April 2017.

39. J. Redmon, S. Divvala, R. Girshick and A. Farhadi, "You only look once: Unified, real-time object detection." In *Proceedings of the IEEE Conference on Computer Vision and Pattern Recognition*, pp. 779–788, 2016, preprint https://arxiv.org/abs/1506.02640

40. O. A. B. Penatti, K. Nogueira and J. A. Dos Santos, "Do deep features generalize from everyday objects to remote sensing and aerial scenes domains?" In *Proceedings of the IEEE Conference on Computer Vision and Pattern Recognition Workshops (CVPRW) 2015*, pp. 44–51, June 2015.

41. F. Hu, G.-S. Xia, J. Hu and L. Zhang, "Transferring deep convolutional neural networks for the scene classification of high-resolution remote sensing imagery." *Remote Sens.*, 7(11), pp. 14680–14707, 2015.

42. A. Gangopadhyay, S. M. Tripathi, I. Jindal and S. Raman, "SA-CNN: Dynamic scene classification using convolutional neural networks." 2015, preprint https://arxiv.org/abs/1502.05243.

43. C. Hung, Z. Xu and S. Sukkarieh, "Feature learning based approach for weed classification using high resolution aerial images from a digital camera mounted on a UAV." *Remote Sens.*, 6(12), pp. 12037–12054, 2014.

44. A. Bry, A. Bachrach and N. Roy, "State estimation for aggressive flight in GPS-denied environments using onboard sensing." In *2012 IEEE International Conference on Robotics and Automation (ICRA)*, IEEE, pp. 1–8, 2012.

45. T. Xiang, F. Jiang, G. Lan, J. Sun, G. Liu, Q. Hao and C. Wang, "UAV based target tracking and recognition." *IEEE Int Conf Multisens Fusion Integr Intell Syst*, pp. 400–405, 2017. doi:10.1109/MFI.2016.7849521

46. A. Guillen-Perez, R. Sanchez-Iborra, M. D. Cano, J. C. Sanchez-Aarnoutse and J. GarciaHaro, "Wi-Fi networks on drones." In *ITU Kaleidoscope: ICTs for a Sustainable World (ITU WT)*, pp. 1–8, 2016.

47. E. Ferro and F. Potorti, "Bluetooth and Wi-Fi wireless protocols: A survey and a comparison." *IEEE Wirel Commun*, 12(1), pp. 12–26, 2005.

48. G. V. Crosby and F. Vafa, "Wireless sensor networks and LTE-A network convergence." In *38th IEEE Annual Conference on Local Computer Networks*, pp. 731–734, 2013.

49. S. Qazi, A. S. Siddiqui and A. I. Wagan, "UAV based real time video surveillance over 4G LTE." In *IEEE International Conference on Open-Source Systems & Technologies (ICOSST)*, pp. 141–145, 2015.

50. I. Dalmasso, I. Galletti, R. Giuliano and F. Mazzenga, "WiMAX networks for emergency management based on UAVs." In *1st IEEE AESS European Conference on Satellite Telecommunications (ESTEL)*, pp. 1–6, 2012.

51. P. Kinney, "Zigbee technology: Wireless control that simply works." In *Communications Design Conference*, vol. 2, pp. 1–7, 2003.

52. J. de Carvalho Silva, J. J. Rodrigues, A. M. Alberti, P. Solic and A. L. Aquino, "LoRaWAN—A low power WAN protocol for Internet of Things: A review and opportunities." In *2nd IEEE International Multidisciplinary Conference on Computer and Energy Science (SpliTech)*, pp. 1–6, 2017.

53. Q. Wu, G. Y. Li, W. Chen, D. W. K. Ng and R. Schober, "An overview of sustainable green 5G networks." *IEEE Wirel Commun*, 24(4), pp. 72–80, 2017.

54. M. Bacco, A. Berton, A. Gotta and L. Caviglione, "IEEE 802.15. 4 Air-Ground UAV communications in smart farming scenarios." *IEEE Commun Lett*, 22(9), pp. 1910–1913, 2018.

55. M. D. R.-Moreno, D. Castaño, D. F. Barrero and A. M. Hellín, "Efficient services management in libraries using AI and wireless techniques." *Expert Syst Appl*, 41(17), pp. 7904–7913, 2014.

56. A. Fouda, A. S. Ibrahim, I. Guvenc and M. Ghosh, "UAV-based in-band integrated access and backhaul for 5G communications." In *88th IEEE Vehicular Technology Conference (VTC-Fall)*, pp. 1–5, 2019.

57. U. Challita, A. Ferdowsi, M. Chen and W. Saad, "Machine learning for wireless connectivity and security of cellular-connected UAVs." *IEEE Wirel Commun*, 26(1), pp. 28–35, 2019.

58. J. L. Junell, E.-J. VanKampen, C. de Visser and Q. Chu, "Reinforcement learning applied to a quadrotor guidance law in autonomous flight." In *AIAA Guidance, Navigation, and Control Conference*, AIAA 2015-1990, 1–12, 2015.

59. B. Zhang, W. Liu, Z. Mao, J. Liu and L. Shen, "Cooperative and geometric learning algorithm (CGLA) for path planning of UAVs with limited information." *Automatica*, 50(3), pp. 809–820, 2014.

60. F. Aznar, M. Pujol and R. Rizo, "Visual navigation for UAV with map references using ConvNets." In *Advances in Artificial Intelligence*, vol. 9868 of Lecture Notes in Computer Science, Springer, pp. 13–22, 2016.

61. P. T. Jardine, S. Givigi and S. Yousefi, "Parameter tuning for prediction-based quadcopter trajectory planning using learning automata." *IFAC-PapersOnLine*, 50(1), pp. 2341–2346, 2017.

62. G. Chowdhary, T. Wu, M. Cutler and J. P. How, "Rapid transfer of controllers between UAVs using learning-based adaptive control." In *2013 IEEE International Conference on Robotics and Automation (ICRA)*, IEEE, pp. 5409–5416, 2013.

63. S. Bansal, A. K. Akametalu, F. J. Jiang, F. Laine and C. J. Tomlin, "Learning quadrotor dynamics using neural network for flight control." In *2016 IEEE 55th Conference on Decision and Control (CDC)*, IEEE, pp. 4653–4660, 2016.

64. M. P. Wellman, W. E. Walsh, P. R. Wurman and J. K. MacKie-Mason, "Auction protocols for decentralized scheduling." *Games Econ Behav*, 35, pp. 271–303, 2001.

65. S. Park, L. Zhang and S. Chakraborty, "Battery assignment and scheduling for drone delivery businesses." In *Proceedings IEEE/ACM International Symposium on Low Power Electronics and Design*, pp. 1–6, 2017.

66. M. Shin, J. Kim and M. Levorato, "Auction-based charging scheduling with deep learning framework for multi-drone networks." In *IEEE Transactions on Vehicular Technology*, vol. 68, no. 5, pp. 4235–4248, May 2019.

67. C.-Y. Lee, "Cooperative drone positioning measuring in Internet-of-Drones." *IEEE 17th Annual Consumer Communications & Networking Conference (CCNC)*, 1–3, 2020.

68. A. Bachrach, S. Prentice, R. He, P. Henry, A. S. Huang, M. Krainin, D. Maturana, D. Fox and N. Roy, "Estimation, planning, and mapping for autonomous flight using an RGBD camera in GPS-denied environments." *Int J Rob Res*, 31(11), pp. 1320–1343, 2012.

69. H. Zhang, G. Wang, Z. Lei and J.-N. Hwang, "Eye in the sky: Drone-based object tracking and 3D localization." In *Proceedings of the 27th ACM International Conference on Multimedia*, 9 pages, 2019.

70. F. Fraundorfer, L. Heng, D. Honegger, G. H. Lee, L. Meier, P. Tanskanen and M. Pollefeys, "Vision-based autonomous mapping and exploration using a quadrotor mav." In *2012 IEEE/RSJ International Conference on ADVANCED ROBOTICS 11 Intelligent Robots and Systems (IROS)*, IEEE, pp. 4557–4564, 2012.

71. M. Rabah, A. Rohan, M. Haghbayan, J. Plosila and S. Kim, "Heterogeneous parallelization for object detection and tracking in UAVs," In *IEEE Access*, vol. 8, pp. 42784–42793, 2020, doi:10.1109/ACCESS.2020.2977120.

72. X. Guo, S. Denman, C. Fookes, L. Mejias and S. Sridharan, "Automatic UAV forced landing site detection using machine learning." In *2014 International Conference on Digital Image Computing: Techniques and Applications (DICTA)*, IEEE, pp. 1–7, 2014.

73. Y. LeCun, Y. Bengio and G. Hinton, "Deep learning." *Nature*, 521, pp. 436–444, 2015.

74. A. Kouris and C. Bouganis, "Learning to fly by myself: A self-supervised CNN-based approach for autonomous navigation." In *Proceedings of the 2018 IEEE/RSJ International Conference on Intelligent Robots and Systems (IROS)*, Madrid, Spain, pp. 1–9, 1–5 October 2018.

75. D. Gandhi, L. Pinto and A. Gupta, "Learning to fly by crashing." *IEEE/RSJ Int Conf Intell Robot Syst (IROS)*, 2017, pp. 3948–3955, 2017.

76. S. Y. Choi and C. Dowan, "Unmanned aerial vehicles using machine learning for autonomous flight; state-of-the-art." *Advanced Robotics*, 33(6), pp. 265–277, 2019.

77. A. Giusti, J. Guzzi, D. C. Cireşan, F.-L. He, J. P. Rodríguez, F. Fontana, M. Faessler, C. Forster, J. Schmidhuber, G. Di Caro, D. Scaramuzza and L. M. Gambardella, "A machine learning approach to visual perception of forest trails for mobile robots." In *IEEE Robotics and Automation Letters*, vol. 1, no. 2, pp. 661–667, July 2016, doi:10.1109/LRA.2015.2509024.

78. R. Barták and M. Vomlelová, "Using machine learning to identify activities of a flying drone from sensor readings." In *(Proceedings of) the Thirtieth International Florida Artificial Intelligence Research Society Conference*, FLAIRS, pp. 436–441, 2017.

79. M. Karimibiuki, M. Aibin, Y. Lai, R. Khan, R. Norfield and A. Hunter, "Drones' face off: Authentication by machine learning in autonomous IoT systems." *IEEE Ubiquitous Computing, Electronics & Mobile Communication Conference*, 329–333, 2019.

80. M. Das and S. K. Ghosh, "A deep-learning-based forecasting ensemble to predict missing data for remotesensing analysis." *IEEE J Sel Top Appl Earth Observ Remote Sens*, 10, pp. 5228–5236, 2017.

81. S. Al-Emadi, A. Al-Ali, A. Mohammad and A. Al-Ali, "Audio based drone detection and identification using deep learning." In *Proceedings of the 2019 15th International Wireless Communications & Mobile Computing Conference (IWCMC)*, Tangier, Morocco, pp. 459–464, 24–28 June 2019.

82. J. Ye, C. Zhang, H. Lei, G. Pan and Z. Ding, "Secure UAV-to-UAV systems with spatially random UAVs." In *IEEE Wireless Communications Letters*, vol. 8, no. 2, pp. 564–567, April 2019, doi: 10.1109/LWC.2018.2879842.

83. U. Challita, W. Saad and C. Bettstetter, "Interference management for cellular-connected UAVs: A deep reinforcement learning approach." In *IEEE Transactions on Wireless Communications*, vol. 18, no. 4, pp. 2125–2140, April 2019, doi: 10.1109/TWC.2019.2900035.

84. K. Li, W. Ni and E. Tovar, "On-board deep Q-network for UAV-assisted online power transfer and data collection." 2019, arXiv preprint arXiv:1906.07064.
85. X. Liu, Y. Liu, Y. Chen and L. Hanzo, "Trajectory design and power control for Multi-UAV assisted wireless networks: A machine learning approach." In *IEEE Transactions on Vehicular Technology*, vol. 68, no. 8, pp. 7957–7969, August 2019.
86. K. Kumar, S. Kumar, O. P. Kaiwartya, P. K. Kashyap, J. Lloret and H. song, "Drone assisted flying ad-hoc networks: Mobility and service oriented modeling using neuro-fuzzy." *Ad Hoc Networks*, vol. 106, 102242, 1 September 2020.
87. M. M. Kurdi, A. K. Dadykin, I. Elzein and I. S. Ahmad, "Proposed system of artificial Neural Network for positioning and navigation of UAV-UGV." *2018 Electric Electronics, Computer Science, Biomedical Engineerings' Meeting (EBBT)*, Istanbul, 2018.
88. J. Zhang, J. Yan, P. Zhang and X. Kong, "Design and information architectures for an unmanned aerial vehicle cooperative formation tracking controller." *IEEE Access*, 6, pp. 45821–45833, 2018.
89. L. M. Dang, S. I. Hassan, I. Suhyeon, A. K. Sangaiah, I. Mehmood, S. Rho, S. Seo and H. Moon, "UAV based wilt detection system via convolutional neural networks." *Sustain Comput Informat Syst*, 28, pp. 100250, 2020.
90. A. Ghaderi and V. Athitsos, "Selective unsupervised feature learning with convolutional neural network (S-CNN)." In *Proceedings of the 2016 23rd International Conference on Pattern Recognition (ICPR)*, IEEE, pp. 2486–2490, December 2016.
91. J. Kober, J. A. Bagnell and J. Peters, "Reinforcement learning in robotics: A survey." *International Journal of Robotics Research*, 32(11), pp. 1238–1274, 2013.
92. T. M. Hoang, N. M. Nguyen and T. Q. Duong, "Detection of eavesdropping attack in UAV-aided wireless systems: Unsupervised learning with one-class SVM and K-means clustering." In *IEEE Wireless Communications Letters*, vol. 9, no. 2, pp. 139–142, February 2020, doi: 10.1109/LWC.2019.2945022.
93. X. Yue, Y. Liu, J. Wang, H. Song and H, Cao, "Software defined radio and wireless acoustic networking for amateur drone surveillance." *IEEE Commun Mag*, 56, pp. 90–97, 2018. doi: 10.1109/MCOM.2018.1700423.
94. S. Wang, F. Jiang, B. Zhang, R. Ma and Q. Hao, "Development of UAV-based target tracking and recognition systems." In *IEEE Transactions on Intelligent Transportation Systems*, pp. 1–19, 2019.
95. V. Sharma, D. N. K. Jayakody and K. Srinivasan, "On the positioning likelihood of UAVs in 5G networks." *Physical Comm.*, 31, pp. 1–9, 2018.
96. Z. Ma, C. Wang, Y. Niu, X. Wang and L. Shen, "A saliency-based reinforcement learning approach for a UAV to avoid flying obstacles." *Robotics and Autonomous Syst.*, 100, pp. 108–118, 2018.
97. M. S. Pillai, G. Chaudhary, M. Khari and R. González Crespo, "Real-time automatic automobile accident detection through CCTV using deep learning." *Soft Computing*, Springer, 2020.
98. V. Srivastava, S. Srivastava, G. Chaudhary and F. Al-Turjman, "A systematic approach for COVID-19 predictions and parameter estimation." *Personal and Ubiquitous Compt.*, pp. 1–13, 2020.
99. H. M. R. Afzal, S. Luo, M. K. Afzal, G. Chaudhary, M. Khari and S. A. P. Kumar, "3D face reconstruction from single 2D image using distinctive features." *IEEE Access*, 8, pp. 180681–180689, 2020.

3 Digital Twin in Agriculture Sector
Detection of Disease Using Deep Learning

Meenu Gupta and M. Kumari
Chandigarh University

C. Ved
Bharati Vidyapeeth's College of Engineering

CONTENTS

3.1 INTRODUCTION: BACKGROUND AND DRIVING FORCES

Agricultural plants are intimidated by the occurrence of diseases and pests, which affect their production. The loss of world production because of plant diseases is

DOI: 10.1201/9781003132868-3

estimated to be approximately 10%. Plant diseases are primarily caused by pathogenic agents such as bacteria, fungi, and viruses. Conventionally, the epidemics in plants are diagnosed by experts by visual examination, which is susceptible to errors [1]. Science and technology advances are supplying more delicate techniques and tools for initial detection, control, and prevention of diseases and pests [2]. The latest revolution in digital cameras and smartphones are covering a path to use these economic image acquisition instruments or automate device integration for plant disease identification applications. It is essential to detect the diseases in plants on time and give advice to farmers who want a suggestion for enhancing crops (or yield). The occurrence of plant diseases depends on specified epidemiological and environmental factors, thus often have an uneven field distribution. This book chapter provides a comprehensive overview of traditional methods, the latest trends, and advances in crop disease detection with sensors. It presents recent advances in the use of plant image sensors for detection, recognition, and evaluation of plant diseases on different scales. Multi-spectral, optical, gravimetric, conductivity, hyperspectral, and thermography sensors are the most promising plant sensor types that are used for monitoring plant diseases. A plant is considered diseased when any abiotic or biotic factor negatively affects its growth and development. The disease can be chronic, acute, and can accelerate at a rapid pace. It can be expressed in many ways. The symptoms include leaf curling, morphological changes, chlorosis, premature abscission of plant, and change in leaf angle, stunting, or wilting. Rapid and accurate quantification and detection of early symptoms are usually tricky. Thus, there is a need for lots of research to estimate the losses occurring due to the prevalence of diseases in different and changed environmental conditions and locations. In the agricultural field, pets and diseases cause enormous economic damage to farmers due to reduced yields and high cost of pesticides. For this, the various recent sensor networks and their usability techniques have shown a high potential in diagnosing diseases and in-plant monitoring for areas with infected plants. These techniques attract significant attention in the agricultural field, where disease detection is the most critical work. Various reviews have been done for recognizing foliage diseases, utilizing image processing schemes and machine learning algorithms in different plants. It is shown that imagining sensors can be utilized to analyze and detect plant diseases in many ways, such as finding specific properties of diseased and healthy plants—and thus the quality of plants—by measuring the amount of radiation reflected from their leaves and canopies [3].

3.2 LITERATURE REVIEW

The latest literature survey of disease detection in crops is as follows: In [4], the current developments in RS systems, their features, feature selection methods, conducting a comprehensive model, monitoring algorithms, and plants effective monitoring have been summarized. Here, it is expected that those diseases that can be remotely sensed and monitored can lead to research achievements and provoke novelty thoughts, along with encouraging corresponding techniques, methods, and theories, not only in production, but also in practice. From the literature review, it is clear that there is a lack of information on using sensors for crop disease detection. Thus, this book chapter highlights primary sensors used in crop disease detection.

In [5], a decision support system for Late Blight disease using integrated data for treatment of the disease has been presented. Here, a sensor system and a cloud server to collect the data about humidity and temperature have been developed. Using an SMS, farmers can be notified of the initial attack of "Late Blight" disease. The application can be enhanced by an emerging IoT platform that stores plant's images and investigates the unexpected transformations on the plants' leaves in the form of pale green or pale brown. In the study in [6], *Terminalia bellerica* inhibited the QS-controlled virulence and minimize biofilm formation in *Pseudomonas aeruginosa*. It is presented that the extract has reduced the biofilm production in *P. aeruginosa*. Thus, *T. bellerica*'s antibiofilm and QS inhibition features may be functional in conventional antibiotics to react with the bacterial infections. Also, studies are required to identify the pharmacokinetic properties of *T. bellerica* to utilize their clinical value.

Further, in [7], the authors have designed a hybrid method for detection of crop leaf diseases using Convolutional Neural Network (CNN) and autoencoder combination. It provides a new approach to identify crop diseases, using convoluted encoder networks, having crop leaf images. The obtained results show that convolution filters like 2×2 and 3×3 are utilized to determine leaf images. It also attained accuracy variations for a distinct number of epochs and different convolution filter size [8,9]. Thus, with the help of a 2×2 convolution filter and a 3×3 filter size, the accuracy can be achieved up to 97.50% and 100%, respectively.

In this paper, the image transmission system concept and theory have been discussed. Also, image information of diseases and insects, using a transmission framework system, by adopting an image compression standard has been designed. The results show that damaging image information compression can enhance data transmission quality and efficiency, while reducing network bandwidth [10]. In [11], the hyperspectral sensors and imaging schemes for incompatible plant- pathogen and compatible interactions have been studied. It includes intuitive and smart solutions for plant features, assessing in plant phenotyping and making plant protection measures' decisions with precision in an agricultural context. In [3], a technique to analyze crop pests, on the basis of UAV remote sensing is proposed and investigated for detecting agricultural crop pests in growing season. Here, a SIFT algorithm is utilized for image mosaic and matching with good results.

In [12], a review of the wireless sensor network, its applications, and its significance in precision farming for enhancing agriculture has been presented. It has been stated that it can reduce the cost, improve the lifespan, and help in recovering the monetary loss. In [13], a technique incorporating the minimum noise fraction (MNF) transformation, pure pixels endmember selection, multi-dimensional visualization, and spectral angle mapping (SAM) has been developed for hyperspectral image processing of diseased tomato crops. The results show that the first 28 signal eigen images can be utilized for classification of the image pixels, visualization, and SAM, while the rest are noise-dominated. In [6], the basic principles of hyperspectral sensors for plant protection and disease detection have been presented. It shows that it is feasible to detect the disease before visible symptoms. It also helps in disease quantification and identification in spectral signature. In [14], several artificial neural network (ANN) methods to segment the plant's disease are presented. It is shown that these methods help in back propagation algorithms, self-organizing feature maps, SVMs and many others to classify, as well as identify several plants' diseases, utilizing image processing methods.

3.3 DETECTION METHODS OF CROP DISEASES

3.3.1 METHOD OF DISEASE DETECTION

Commercially, visual estimates find a disease based on the features of plant disease symptoms such as tumors, cankers, blight, galls, lesions, and damping-off with visible indications of a pathogen such as *Pucciniales urediniospores*, *Erysiphales conidia*, and mycelium [15]. However, visual estimation is executed by trained experts and has been an intensive research and investigation subject [16].

3.3.2 PRINCIPLES IN MONITORING PLANT DISEASES

In observing the host interactions and pathogen, various plant diseases cause different types of plant damages and symptoms. It should be eminent that not all plant diseases are appropriate for image-sensed detection, as some of them are deficient in identifiable properties [17]. However, some plants borne root and soil diseases that seed systemic effects on the plant's physiology, which can be detected. Thus, an essential need in monitoring and detecting plant diseases via image sense belongs to a specific response presence that is capable of being recognized by a definite sensors system [4,18]. Figure 3.1 shows different types of plant diseases.

Out of the plants' physiological changes and symptoms that are caused by diseases, there are four kinds of damages:

1. Biomass reduction and LAI decrement: This kind of destruction happens in pest attacks.
2. Pustules or lesions due to infection: These are necrotic or sori tissues that occur as common symptoms of diseases. They differ in their shapes and colors between diseases.
3. Pigment systems destruction: In many cases, infection of plants by a disease can cause organelles or chloroplast destruction, causing variations in pigment contents.
4. Wilting: The rigidity loss because of dehydration is an uncommon plant disease symptom, and it can be simply confused with drought stress. In some infected situations, water flow can be blocked by the vascular system in plants, which causes plant's dehydration [19].

It should be eminent that plant disease occurrence can cause multiple forms of felling, as stated above. Imaged sensing monitoring should be capable of catching

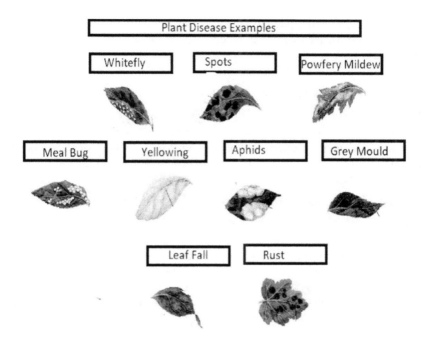

FIGURE 3.1   Different types of plant diseases [1].

symptoms accumulation. The disease infection attack behaves as a temporal process. Various forms of interaction superimpose with each other and show different stages of severities. Without proper control, the harmed photosynthetic system will conduct biomass decrease and will hinder plant's water metabolism [20]. Thus, various infected plants show symptoms, particularly, wilting in the last stage. This will complicate monitoring and detection of some diseases. Therefore, it is essential to find specific RS features at distinct development disease stages. The distinctive temporal patterns' impact on symptoms also allows temporal characteristics extraction to minimize monitoring uncertainty [4]. On the whole, plant disease occurrence depends on specific environmental conditions. For the diseases that exhibit distributed fields, optical sensing schemes are useful in finding main disease areas and foci differing in disease fields' severity. In compound with advanced techniques of data analysis, these methods can be utilized for targeted pest management in feasible crop production. Targeted and site-specific pesticide applications, as per precision crop secure strategies, result in reduction in the use of pesticide and thus, minimize ecological impact and economic expense in agricultural plant production systems [21]. Crop phenotyping evaluates the performance and appearance of a genotype under different environmental factors. During the plant breeding procedure, a large number of distinct genotypes are examined for abiotic stress resistance, disease, produce quality, yield, and various secondary traits [22]. The host-pathogen susceptibility and interactions of breeding material should be evaluated efficiently in disease resistance breeding, as plant phenotyping is not only time-consuming but also labor-intensive. Currently, phenotyping was utilized as a replica for sensor-based analysis and noninvasive imaging of biochemical plant, anatomical, and physiological properties.

3.3.3 LATEST DISEASE DETECTION METHODS

In-depth research has currently identified novel schemes for detection based on the sensor for identification, screening, and plant disease quantification. The sensors for plants assess the imaged properties of plants within distinct areas of electromagnetic spectrum and are capable of using data beyond the visible limit [23].

3.3.4 USE OF DEEP LEARNING TO PREDICT/CLASSIFY DISEASE IN CROPS

In Figure 3.2, the process gets started from dataset extraction from captured images [24]. Initially, dataset goes through a data cleaning process to remove redundant or noisy images. Then, the data splitting process takes place in which 80% data is used for the training set and 20% will be used for the validation set. Deep learning models get trained from scratch by using transfer learning process. Training and validation points get plotted after developing a deep learning model for the purpose of indicating significance [25]. For classification of plant diseases, we can determine the parameters of performance metrics such as F-1 scoring and accuracy, and finally, visualization technique is used to detect localization points in images.

3.3.5 IMPACT OF MODERN OPTICAL SENSORS ON DETECTION

The different sensors are used for the diagnosis and monitoring of plant diseases as the conventional methods are the "gold standard" in plant disease detection, which depends on culture-based symptoms, biochemical identifications, and morphological observations. They allow detection of initial changes in plant physiology because of biotic stress, as the disease can produce modifications in leaf shape, tissue color, transpiration rate, plant density, interaction of solar radiation variations, and canopy morphology with plants [26]. Now the most favorable technique is using sensors that measure temperature, reflectance, and fluorescence. Most of these thermal and spectral sensors were developed for earth remote sensing, military, aircraft, satellite, or industry [27]. Figure 3.3 shows the crop disease detection through two methods, namely, microscopic evaluation and macroscopic evaluation

Analysis using Deep Learning to predict/classify diesease in crops

FIGURE 3.2 Disease prediction and classification using deep learning.

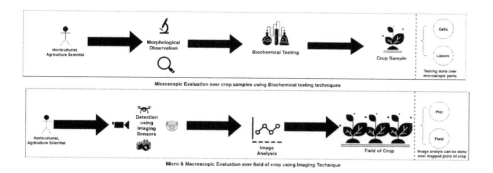

FIGURE 3.3   Crop disease detection techniques.

TABLE 3.1
Comparison between Microscopic and Macroscopic Evaluation Techniques

Serial No.	Microscopic	Macroscopic
1	The disease is detected by visual or morphological observation technique over samples of crop	An imaging technique detects the disease with different types of sensors on the whole plot of the crop
2	Detection accuracy depends upon the experience of agriculture scientist	Detection accuracy depends upon the data analysis & extraction technique
3	Not capable of detecting the whole plot of crop together	Capable of detecting whole plot of crop together, through aerial or land surveillance
4	By biochemical methods over cells & leaves, the disease is detected	By segmentation technique over the leaves, the disease is detected

technique. The fundamental difference between these two techniques is given in Table 3.1.

3.4   STRATEGIES USED FOR CROP DISEASES

3.4.1   Optical Sensors

Optical sensors are favorable tools for noninvasive disease diagnosis and detection. There are various noninvasive and imaging sensors that can be used to diagnose and screen plant diseases. Optical sensor data can be used for efficient detection and diagnosis of diseases. Advanced statistical methods and data analysis are essential factors that are used with various methods to analyze a plant disease detection data, composed by an optical sensor. It includes the following steps: [28]

1. Detection of disease at the initial stage
2. Differentiation of different diseases

3. Abiotic stress causes separation of disease
4. Quantification of disease severity

A visual plant disease approximation method is used for detection and diagnoses of plant diseases, using microscopic evaluation, human raters, etc. Digital cameras are simple to hold and are an easy RGB (red, green, and blue) source of digital images for quantification, identification, and detection. The technical variables of these are easy to use, handheld devices like photosensor light sensitivity, optical focus, digital focus, and spatial resolution, which have enhanced significantly over the years [29]. These sensors are utilized on every stage of determination for detecting and monitoring plants in the growing season. Various researchers have used pattern recognition schemes and machine learning to find plant diseases through RGB images. Also, systematic selection of corresponding properties from the RGB images, enhances classification accuracies. This is because digital image analysis is the most-established technology utilized in plant disease assessment [30]. Figure 3.4 shows the general method of disease prediction and classification.

3.4.2 THERMAL SENSORS

The crop stands microclimate and transpiration changes because of initial plant pathogens infections. Infrared and thermographic cameras can identify this by releasing infrared radiation in the 8–12 mm thermal infrared range along with false-color images illustration [31]. Here each pixel is utilized at distinct spatial and temporal scales from small-scale to airborne applications. It is generally sensitive to environmental conditions like ambient temperature, wind speed, rainfall, and sunlight. The leaf temperature presents a plant transpiration close correlation, which is impacted by pathogen diversity in distinct ways. While lots of foliar pathogens like leaf spots or rusts consist of well-defined and local changes, impairment by root pathogens or systemic infections generally impacts the water flow and the transpiration rate of the whole plant. Regional temperature variations because of disease

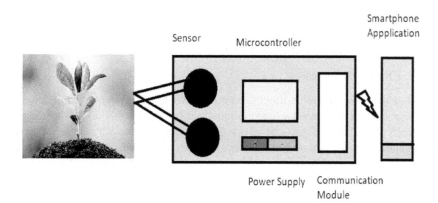

FIGURE 3.4   Disease prediction and classification [3].

or pathogen to defense methods have been reported for plant-virus interchange in tobacco [32].

3.4.3 HYPERSPECTRAL IMAGING IN PLANT

Hyperspectral imaging spectroscopy was observed in the 1980s [15]. In plant science, image sensing is a technique for obtaining data from plants without invasive manipulation or direct contact [33]. The idea has been currently growing by usage of small-scale sensing and proximal or close-range sensing of the plant material. These installed sensors on multiple platforms can be located at strategic points [34].

3.4.4 MULTI-SPECTRAL IMAGING CAMERAS

It may supply data in RGB wavebands and infrared band. The evolution of current hyperspectral sensors enhanced the measured data complexity by a 350–2,500 nm spectral range, with a feasible narrow spectral resolution (<= 1 nm) (Steiner et al. 2008). Opposing non-imaging sensors, which averages the spectral information over a specific area, hyperspectral imaging sensors supply spatial and spectral data for the imaged object [35]. Here, the hyperspectral data can be examined as huge matrices having spatial x, spatial y, and spectral data as reflectance intensity in 3D per waveband [34]. The spatial resolution entirely depends on the object and sensor distance. Also, the spatial resolution has a profound impact on plant disease detection interactions. Airborne sensors are appropriate for the detection of field patches with soil-borne pathogens or in terminal disease stages [38].

3.4.5 GRAVIMETRIC SENSORS

There are two kinds of gravimetric sensors, viz. quartz crystal microbalance (QCM) and surface acoustic wave (SAW) sensors. SAW sensors supply a surface wave that moves along the sensor's surface, while QCM sensors provide a stream by walking through the sensor bulk. The working principle of Gravimetric Sensors consists of piezoelectric sensor coating mass change because of gas absorption, which causes a variation in the frequency, resonant upon exposure to VOCs. SAW sensors include a piezoelectric substrate having an interdigital input gaining electrode and transmitting electrode placed at the substrate surface top. A sensitive thin film is set at interdigital electrodes [36]. Odor molecules communicate with the change in the mass and sensing film unit, which leads to a variation in the frequency. QCM sensors have the same principle to SAW but consist of a distinct device structure. This sensor consists of a quartz chip covered with a gripping sensing membrane, and gold electrodes set attached to the base of the disk. It shows promising performance for observing food quality. The selectivity and sensitivity of these sensors depend on the sensing material type and odor and film compound interaction. Enhancing the sensitivity of these sensors depends on developing explicit sensing materials. The merits of QCM and SAW sensors include small size, low cost, and high sensitivity. Their demerits include short life span and complex fabrication process [37].

3.4.6 CONDUCTIVITY SENSORS

Figure 3.5 shows steps in prediction of plant diseases. These sensors are based on conducting polymer (CP) and metal oxide semiconductor (MOS), which work on variations in conductivity and resistance upon exposure principle. However, the mechanisms responsible are distinct. The sensing materials, conductive sensors substrate, and electrodes are primarily the same.

MOS sensors need a heater. Conducting polymers have lots of advantages over materials. Sensors developed from conducting polymers can work at room temperature. It is a significant merit for portable battery-powered systems as extra heater increases power consumption and minimizes the battery life. Most importantly, in case of array sensors, high discrimination can be attained by utilizing distinct conducting materials, as lots of categories of conducting polymers are present. Although a primary demerit of conducting polymer is aging, it also causes poor performance and sensor drift [37].

3.5 CHALLENGES

The impact of environmental temperature and humidity is a significant challenge in detection of plant diseases as sensor systems are sensitive to the weather of environment. This causes sensor response drift, which minimizes detection signal-to-noise ratio (S/N) of the target. Moreover, humidity reduces the life span of sensors, thus restricting their applications in high-frequency monitoring and assessment of long-term status of plants [3]. Besides the inspired progress that has been attained in

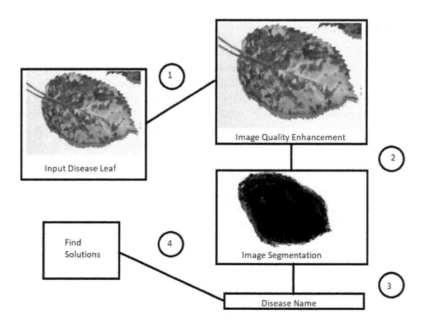

FIGURE 3.5   Disease prediction steps [12].

TABLE 3.2
Challenges in Monitoring Plant Diseases

Issues	Challenges	Trends
Detection of plant diseases at the initial stage	Various plant diseases are symptomless or have general symptoms at the initial stage. The early symptoms of diseases tend to happen at mid- to base layers	Utilize the fluorescence, thermal, LIDAR RS, and SAR feasibility in early symptoms detection, using various multi-angular image sensing
Correctly detect a particular disease under a realistic field state	Various plants stress may happen simultaneously. Some of the crop diseases may show the same symptoms	Explore monitoring models' transferability and features of uniqueness. Begin a base knowledge to help in minimizing uncertainty in the monitoring process.
Regularly analyze the disease's dynamics at a proper resolution	The RS plant system should have enough high resolution at temporal, spectral, and spatial dimensions. Worst weather causes the regular optical images acquisition	Collaborate high- resolution images using Unmanned Aerial Vehicle (UAV) images to generate a time-series RS data through a fusion between radar data and optical RS data [39]
Information and data shaping	Insufficient survey data in monitoring plant diseases. Increase the pooled data accessibility to handle model training and data mining	Set up international projects and allow data collection, experiments, ideas, and models at a global scale

the plant disease monitoring during the last few years, some issues are still present in implementing these techniques. Another challenge is the detection of field conditions. Performance of sensors in real environments requires being reviewed utilizing more extensive field trials. However, the environmental parameters of full and open experiments like humidity, background gas compositions, and temperature are kept changing and uncontrollable. Moreover, background noise produced from other objects in the environment can hide the plants' original properties caused by fungal infection, pest attacks, etc. Thus, a controlled atmosphere that can handle humidity temperature and gas compositions is appropriate for sensor applications. The sensor arrays' development with high selectivity and sensitivity are desirable [38]. Some of the challenges are given in Table 3.2.

3.6 CONCLUSION AND FUTURE SCOPE

Disease detection in plants plays an important role in the field of agriculture to protect the crops from diseases at initial stages. Thus, preventive measures can be taken initially, which will be economically viable for farmers as well as the agriculture sector and will also avoid food wastage. The recent digital twin concept in agriculture offers vertical framing by IoT installation development to enhance the food supply problematic situation. It also allows significant energy savings practices. It is highly required to overcome the challenges in the development of advanced imaging methods of disease detection and prediction to achieve more accuracy in large land areas of crops. Efficiency can be attained by predicting the root cause of the problem by integrating different techniques of deep learning and machine learning. Useful mapping and prediction can be made in plants for identification of diseases based on multiple sensing and imaging parameters.

REFERENCES

1. Navulur, S., Sastry, A.S.C.S., & Giri Prasad, M. N. (2017) "Agricultural Management through Wireless Sensors and Internet of Things" *International Journal of Electrical and Computer Engineering (IJECE)*, *7*(6), 3492–3499.
2. Cui, S., Ling, P., Zhu, H., & Keener, H. M. (2018). Plant pest detection using an artificial nose system: a review. *Sensors*, *18*(2), 378.
3. Yue, J., Lei, T., Li, C., & Zhu, J. (2012). The application of unmanned aerial vehicle remote sensing in quickly monitoring crop pests. *Intelligent Automation & Soft Computing*, *18*(8), 1043–1052.
4. Puthren, P. R., Agrawal, A., & Padma, U. (2018, May). Automated Glaucoma Detection Using Global Statistical Parameters of Retina Fundus Images. In *International Conference on ISMAC in Computational Vision and Bio-Engineering* (pp. 377–388). Springer, Cham. Pandian D., Fernando X., Baig Z., Shi F. (eds)
5. Masood, R., Khan, S. A., & Khan, M. N. A. (2016). Plants disease segmentation using image processing. *International Journal of Modern Education and Computer Science*, *8*(1), 24.
6. Thomas, S., Kuska, M. T., Bohnenkamp, D., Brugger, A., Alisaac, E., Wahabzada, M., … Mahlein, A. K. (2018). Benefits of hyperspectral imaging for plant disease detection and plant protection: a technical perspective. *Journal of Plant Diseases and Protection*, *125*(1), 5–20.
7. Iqbal, Z., Khan, M. A., Sharif, M., Shah, J. H., Rehman, M. H., & Javed, K. (2018). An automated detection and classification of citrus plant diseases using image processing techniques: A review. *Computers and Electronics in Agriculture*, *153*, 12–32.
8. Chaudhary, G., & Srivastava, S. (2020). A robust 2D-Cochlear transform-based palmprint recognition. *Soft Computing*, *24*(3), 2311–2328.
9. Chaudhary, G., Kaur, S., Mehta, B., & Tewani, R. (2019). Observer based fuzzy and PID controlled smart greenhouse. *Journal of Statistics and Management Systems*, *22*(2), 393–401.
10. Natori, T., Ariyama, N., Tsuichihara, S., Takemura, H., & Aikawa, N. (2019, June). Study of Activity Collecting System for Grazing Cattle. In *2019 34th International Technical Conference on Circuits/Systems, Computers and Communications (ITC-CSCC)* (pp. 1–4). IEEE.
11. Kaur, S., Pandey, S., & Goel, S. (2019). Plants disease identification and classification through leaf images: A survey. *Archives of Computational Methods in Engineering*, *26*(2), 507–530.
12. Abd El-kader, S. M., & El-Basioni, B. M. M. (2013). Precision farming solution in Egypt using the wireless sensor network technology. *Egyptian Informatics Journal*, *14*(3), 221–233.
13. Zhang, M., Qin, Z., Liu, X., & Ustin, S. L. (2003). Detection of stress in tomatoes induced by late blight disease in California, USA, using hyperspectral remote sensing. *International Journal of Applied Earth Observation and Geoinformation*, *4*(4), 295–310.
14. Khirade, S. D., & Patil, A. B. (2015, February). Plant disease detection using image processing. In *2015 International conference on computing communication control and automation* (pp. 768–771). IEEE.

15. Newby, Z., Murphy, R. J., Guest, D. I., Ramp, D., & Liew, E. Y. (2019). Detecting symptoms of Phytophthora cinnamomi infection in Australian native vegetation using reflectance spectrometry: Complex effects of water stress and species susceptibility. *Australasian Plant Pathology, 48*(4), 409–424.
16. Foughali, K., Fathallah, K., & Frihida, A. (2018). Using Cloud IOT for disease prevention in precision agriculture. *Procedia Computer Science, 130,* 575–582.
17. Nabi, F., Jamwal, S., & Padmanbh, K. (2020). Wireless sensor network in precision farming for forecasting and monitoring of apple disease: A survey. *International Journal of Information Technology,* 1–12.
18. Usha, K., & Singh, B. (2013). Potential applications of remote sensing in horticulture— A review. *Scientia Horticulturae, 153,* 71–83.
19. Rani, A. S., & Jyothi, S. (2017). A study on hyper spectral remote sensing pest management. *International Journal on Recent and Innovation Trends in Computing and Communication, 5*(6), 497–503.
20. Zhang, J., Huang, Y., Pu, R., Gonzalez-Moreno, P., Yuan, L., Wu, K., & Huang, W. (2019). Monitoring plant diseases and pests through remote sensing technology: A review. *Computers and Electronics in Agriculture, 165,* 104943.
21. Mirzaei, M., Marofi, S., Abbasi, M., Solgi, E., Karimi, R., & Verrelst, J. (2019). Scenario-based discrimination of common grapevine varieties using in-field hyperspectral data in the western of Iran. *International Journal of Applied Earth Observation and Geoinformation, 80,* 26–37.
22. Durmuş, H., Güneş, E. O., & Kırcı, M. (2017, August). Disease detection on the leaves of the tomato plants by using deep learning. In *2017 6th International Conference on Agro-Geoinformatics* (pp. 1–5). IEEE.
23. Ganesh, P. S., & Rai, V. R. (2018). Attenuation of quorum-sensing-dependent virulence factors and biofilm formation by medicinal plants against antibiotic resistant Pseudomonas aeruginosa. *Journal of traditional and complementary medicine, 8*(1), 170–177.
24. Sladojevic, S., Arsenovic, M., Anderla, A., Culibrk, D., & Stefanovic, D. (2016). Deep neural networks based recognition of plant diseases by leaf image classification. *Computational Intelligence and Neuroscience,* 2016. Volume 2016, Article ID 3289801, 11 pages.
25. Mishra, P., Asaari, M. S. M., Herrero-Langreo, A., Lohumi, S., Diezma, B., & Scheunders, P. (2017). Close range hyperspectral imaging of plants: A review. *Biosystems Engineering, 164,* 49–67.
26. Khamparia, A., Saini, G., Gupta, D., Khanna, A., Tiwari, S., & de Albuquerque, V. H. C. (2020). Seasonal crops disease prediction and classification using deep convolutional encoder network. *Circuits, Systems, and Signal Processing, 39*(2), 818–836.
27. Hunt Jr, E. R., & Daughtry, C. S. (2018). What good are unmanned aircraft systems for agricultural remote sensing and precision agriculture? *International Journal of Remote Sensing, 39*(15–16), 5345–5376.
28. Li, J., Zhang, Z., Liu, Y., Yao, C., Song, W., Xu, X., … Zhang, Y. (2019). Effects of micro-sprinkling with different irrigation amount on grain yield and water use efficiency of winter wheat in the North China Plain. *Agricultural Water Management, 224,* 105736.
29. Mahlein, A. K., Kuska, M. T., Behmann, J., Polder, G., & Walter, A. (2018). Hyperspectral sensors and imaging technologies in phytopathology: state of the art. *Annual Review of Phytopathology, 56,* 535–558.
30. Hansen, P. M., & Schjoerring, J. K. (2003). Reflectance measurement of canopy biomass and nitrogen status in wheat crops using normalized difference vegetation indices and partial least squares regression. *Remote sensing of environment, 86*(4), 542–553.
31. Valente, J., Sanz, D., Barrientos, A., Cerro, J. D., Ribeiro, Á., & Rossi, C. (2011). An air-ground wireless sensor network for crop monitoring. *Sensors, 11*(6), 6088–6108.

32. Mistry, I., Tanwar, S., Tyagi, S., & Kumar, N. (2020). Blockchain for 5G-enabled IoT for industrial automation: A systematic review, solutions, and challenges. *Mechanical Systems and Signal Processing, 135,* 106382.
33. Mahlein, A. K. (2016). Plant disease detection by imaging sensors–parallels and specific demands for precision agriculture and plant phenotyping. *Plant disease, 100*(2), 241–251.
34. Bonadies, S., Lefcourt, A., & Gadsden, S. A. (2016, May). A survey of unmanned ground vehicles with applications to agricultural and environmental sensing. In *Autonomous air and ground sensing systems for agricultural optimization and phenotyping* (Vol. 9866, p. 98660Q). International Society for Optics and Photonics.
35. Apan, A., Held, A., Phinn, S., & Markley, J. (2004). Detecting sugarcane 'orange rust' disease using EO-1 Hyperion hyperspectral imagery. *International Journal of Remote Sensing, 25*(2), 489–498.
36. Hunt Jr, E. R., Daughtry, C. S., Stern, A. J., & Russ, A. L. (2019). Linear transects of imagery increase crop monitoring efficiency using fixed-wing unmanned aircraft systems. *Agricultural & Environmental Letters, 4*(1), 190040.
37. Skottrup, P. D., Nicolaisen, M., & Justesen, A. F. (2008). Towards on-site pathogen detection using antibody-based sensors. *Biosensors and Bioelectronics, 24*(3), 339–348.
38. Vishnu, S., & Ranjith Ram, A. (2015). Plant disease detection using leaf pattern: A review. *International Journal of Innovative Science, Engineering & Technology, 2*(6), 774–780.
39. Abdulridha, J., Batuman, O., & Ampatzidis, Y. (2019). UAV-based remote sensing technique to detect citrus canker disease utilizing hyperspectral imaging and machine learning. *Remote Sensing, 11*(11), 1373.

4 Crop Diseases Detection and Prevention Using AI and Machine Learning Techniques

Meenu Gupta, Rakesh Kumar, and Divya Singh
Chandigarh University

CONTENTS

4.1 INTRODUCTION

For around 58% of the population in India, agricultural production is the predominant source of subsistence [1]. India has the world's tenth-largest repository of agricultural production. Including 20 agro-based provinces, India is home to all 15 of the globe's

DOI: 10.1201/9781003132868-4

prime ecosystems. India is the great producer of herbs, grains, dairy, and rayon; and therefore, the second successful supplier of wheat, rice, sugar cane, cotton, and edible oils. Throughout the growing seasons of 2018–2019, like raining, pollution, and drought, agricultural goods resources are estimated even at a record 284.95 million tons. Indian government planned for cultivation of 291.10 million tons of food grain in 2019–2020. As per preliminary estimates, the horticultural crop output of India is projected at 310.7 million metrics tons at a peak high in 2018–2019. India seems to have a huge potential for growth with over 535.78 million livestock, accounting for about 31% of the world's population [2]. Gross value added (GVA) burgeoning in farming and confederated regions reached 4% in Financial Year 2020 [3]. India's GDP from agriculture declined to INR 4,546.58 trillion in the second part of 2020, from INR 5,306.26 trillion in the first part of 2020 [4]. As Kant says [5], if India's GDP has to increase 9%–10% in the next 30 years, this cannot be achieved without annotated revolution in the farming industry [5]. For the year 2020–2021, the Indian government has been able to outpace inflation of 3%, fully in line with NITI Aayog [6]. There are significant challenges to farming systems due to crop disorders, since these significantly decrease the performance and productivity of farming. In agricultural development, premature detection and analysis of diseases is a massive issue. A wide range of crops can be grown by farmers, but crop diseases hinder their production. Different plant species suffer from a variety of illnesses. Viruses, fungi, and bacteria are the main causative agents of crop leaf diseases. Nearly 40% of global crops are recorded to be lost to numerous diseases [7]. Productivity is attacked by unforeseen ailments in the field; hence, recognition of these ailments at an initial point can boost productivity. The researchers have made a variety of automation mechanisms for classification of crop diseases in the discipline of Computer Vision [8]. Approaches such as support vector machine (SVM) classifier [9], image decomposition using K-means clustering [10], and Radial Basis Function Neural Network (RBFNN) [11] were used for feature extraction by most experts. Deep learning (DL) methodology, such as CNN [12], has now become quite common and attracted the attention of researchers. In the past few decades, advances have been made in the field of DL. DL is a machine learning technique that employs an algorithm to scan similarities in large amounts of data, and throughout this event, nearly 50,000 images were given access by Plant Village [13] of diseased plants. This chapter will address the identification and forecasting of crop diseases using the SCYP model that uses deep learning, which collects parameters such as temperature, moisture, sunlight, and weather and also farming sensor data (date of maturation, data on disease, crop condition, temperature of the soil, etc.), identifies crop illness using CNN, and calculates crop productivity based on several factors. The SCYP includes a Crop Disease Diagnosis Module (CDDM) sponsored by CNN to recognize crop diseases via crop photographs, IPM to obtain crop diseases through the Google Vision API and portrait photography, and a Crop Yield Prediction Module (CYPM) to focus on data such as field crops, atmosphere pressure, light, and so on are used in ANN.

4.2 BACKGROUND

Suleiman Bani Ahmad et al. [14] discussed the model-based neural network recognition and identified that the whole model is very beneficial in the identification of leaf diseases, although K-means clustering delivers consistent results as in RGB texture analysis (Red, Green, and Blue). Further in Ref. [15], Zhang et al. addressed a recent three-dimensional diagnostic technique for CNN that instantaneously defines crops using remotely sensed spatiotemporal images. In three-dimensional CNN strategies an active learning method has been proposed for increasing the classification performance to a high level. CNN-based technologies work better than traditional methods. Three-dimensional CNN performs better than 2D CNN; however, SVM has been the most successful amongst traditional methods [16]. Next in Ref. [17], Singh et al. noticed a genetic algorithm, and presented feature extraction and soft virtualization methodology for automated detection and diagnosis of leaf illness. The identification has been initially achieved with a reliability level of only 86.54% that makes use of the minimum gap principle with K-mean clustering. The suggested technique increased this accuracy to 93.63%, which had been further enhanced to 95.71% by using SVM only with the suggested algorithm. The overall performance of the suggested method was 97.6%.

In Ref. [18], an innovative technique to assess the existence of four distinct medical conditions, namely, safe, downy mildew, powdery mildew, and black rot, using one-class classification is demonstrated by Pantazi et al. To recognize four different chronic diseases, the established classifier was tested on a vine leaf surface. The unique elevated importance of the present program is the capacity of generalization, which has been demonstrated by research in a variety of leaf specimens from different plant species. In most of the cases, the data proved that the design was successful. More precisely, 44 of 46 varieties of plant diseases evaluated were correctly demonstrated, giving a 95% overall success rate. Further in Ref. [19], Juan et al. designed a framework for recognizing crop diseases using ESRGAN. By converting LR photos to GAN-supported SR pictures, we created a method for identifying crop disorder on LR photos. Second, to elicit the corresponding SR images, researchers impose ESRGAN on LR images. Researchers use transition learning to fine-tune the learning process Image Net-trained design due to limited crop data. A stable state is reached by our SR model after two fine-tuning moves, and the produced images acquire an attraction effect. Then, we use the triggering SR images to handle disease stratification experiments. The experimental intrinsic part indicates that by implementing the recommended SR model, the identification fidelity can be greatly reinforced; this ensures that our SR model will recreate valuable details for serious crop diseases.

In addition [20], Kadir and Abdul offered the identification of crop species as a vital feature before we define the class of disorder. In their scientific work, Abdul Kadir used color characteristics to calculate mean and standard deviation; the pixel data for leaves are comprised of skewness and kurtosis. They merged factors that distinguish the texture of an image through grey-level co-occurrence matrix (GLCM) functions. It induces a GLCM that tends to produce statistical measurements by determining how often pixel sets of unique values arise in an image and a given partial relationship. Further in Ref. [21], K.C. Kamal et al. explored techniques for fully automatic identification and evaluation of plant diseases at their initial stages involving CNN model and transfer learning (TL). They used simplistic leaf photographs acquired in situ than in comfortable conditions of balanced and unhealthy plants. They proposed a rapid implementation that used a procurement approach of freezing blocks of layers or other specific layers to execute fine-tuning of government prototypes. Six major model architectures were developed, and they are distinct, with a performance of 99.74% achieved through the fine-tuning of DL framework history qualified on Image Net as the main rendering. Further in Ref. [22], Morbekar et al. developed a model using a new methodology to detect plant diseases by an object-tracking system called YOLO (You Only Look Once). YOLO is a diagnostic method for real-time leaf photographs at 45 frames per second, which is higher than other methods for object recognition. In only one analysis, the bounding boxes and class probabilities were projected by just a single classifier. They were using the Plant Village sample throughout this document to address potential diseases. For disease detection, the YOLO DL algorithm is used to achieve higher precision than that of other frameworks. In Ref. [23], Raikar et al. also highlighted that depiction all across the etym of the technological platform's websites, as well as a mobile application-"e-farm for C.R.O.P" but it is of one that assisted agricultural suitably. This included verification of the illness of crop, estimation of crop production, and suggestion of unequalled crop as a premier destination. Multi-lingual support for users, topic assembly, and weather forecasts have been provided in other functions. The website and the smartphone device would fill the gap between technologies and farming. For commercial farms, the planned framework would be very convenient and accessible.

Further in Ref. [24], Yun et al. described the photographs of afflicted crop leaves; quantitatively assessed characteristics such as color, shape, and texture by image processing; and acquired and selected a model for optimizing meteorological characteristics with either the maximum quality or performance probability rate by the method of deterioration of qualities. To evaluate the estimated classification, the PNN classifier was chosen. The outcomes of the analysis of three datasets of diseased leaf images from cucumber showed whether the combined applications of technologies for leaf image analysis, the parasite meteoric data, and PNNs classifier can effectively identify crop diseases, and hence the diagnostic reliability rate was over 90% (Table 4.1).

4.3 CHALLENGES FACED DURING DETECTION OF CROP DISEASES

There are some challenges that farmers face:

- A growing threat to growers is fungicide resistance. As the breakthrough of recent activism, modes remain unknown and farmers are persuaded to hedge on applications for sound strength influence as merchandise emerges as less rapidly priced as widely distributed formulations, and the ability to lose complete fungicide training becomes an ever-growing risk.
- Demographic shifts and the scarcity of cropland. There are fewer persons remain in urban areas who can understand how complicated it is to establish a crop. Along with that, it's necessary to mold a different set of equipment for protecting crops. The decline in the size of farmland would add additional difficulties, and the upcoming residential construction will result in more litigation over the soil and fragrances of the development, as well as the presence of vehicle on major highways.

TABLE 4.1
Comparison of Various Research Works with Outcomes

S. No.	Author	Year	Methodology	Findings/Accuracy
1	Kulkarni and Patil [25]	2012	Filter Gabor and ANN methods were included.	91%.
2	Pujari et al. [26]	2014	IPM related to fungal diseases.	Comparison of IPM related to fungal disease is done.
3	Rastogiet al. [27].	2015	Fuzzy logic is implemented to find the grading and percentile of disease.	Percentile of infection is found and also grading of disease is done.
4	Singh et al. [28]	2015	To determine rice blast disease, the paper included segmentation-based SVM and K-means clustering.	An accuracy of 82% was achieved for classification
5	Sabrol and Satish [29]	2016	Paper classifies six different diseases of tomato plants using image processing techniques. Image processing techniques extract features from images of healthy and diseased plants and later these extracts are classified using classification decision tree.	Accuracy of 97.3% was achieved.
6	Sharma et al. [30]	2017	An automated system for identification of potato crops is implemented. Methods used are adaptive, thresholding for segmentation.	Accuracy of 96% was achieved.
7	Periasamy and Shanmugam [31]	2017	The paper uses remote sensing images for early detection of the crop diseases. Canny edge detection & histogram matching are used.	The detection of disease is done at an early stage.
8	Park et al. [32]	2018	The paper uses CNN-based approach to classify the disease of strawberry plants. The system makes use of DL to diagnose the disease.	An accuracy of 92% is achieved for classification.

- The continuation of a merger focused on the development of new pesticides could have had a significant influence on the performance to identify and promote new products. Either as realized or no longer realized by the considerable majority of the American public, we owe a great deal of gratitude to farming which is now no longer the most powerful. But, in addition, there are specialists working in agencies that carry out experiments to protect our vegetation from dangerous pathogens and avoid large losses.
- There are significant problems with the fate of plant breeding and advancement of genetically modified varieties. There seems to be the quickest demand for better, relatively high, depression types, and a crucial part of the solution must be the role of transgenic technology.
- The danger of converting environment may also put additional sickness and strain on crops [33].

Farmers recognize that they lose vegetation due to farm pathogens as well as parasites, but researchers have discovered that even on an international scale, there is 10%–40% reduction in agricultural productivity of five main crops, in line with documentation of scientists from UC Agriculture and Natural Resources and numerous members of the International Plant Pathology Society. According to the scientists from an international study of crop physical therapy, effects of infectious diseases and agricultural pests, mostly mosquitoes, are minimized in crops such as wheat, rice, corn, soybean, and potato [34].

4.4   A SCYP (SELF-PREDICTABLE CROP YIELD) PLATFORM FOR CROP DISEASE DETECTION

4.4.1   Overview for SCYP

As the arena's populace continues to develop, so does the importance of exactitude or "clever agriculture". Numerous elements, such as meteorological globalization, crop disease, and many more, provide an unprecedented thread to the world's food security. Recently, floor temperature has increased about 0.6° C worldwide. The poor and vulnerable and most food insecure communities on the planet are also the most rationally sound. Ultimately, it is likely that currently precious marine and coastal goods would become even scarcer and more costly, and therefore that deteriorating sanitation and economic resource would make necessary adjustments to a global warming extremely difficult. Crop infections necessitate extra caution when it comes to dietary supplement preservation, but due to the lack of mandatory resources, their expeditious detection continues to remain troublesome in several other factors of the industry [35]. Several more interventions are designed to eradicate the problems of temperature alteration and crop distress. With large numbers turning into a scandal, one more selfish person will do something about it. It has turned out to be obvious that vast data needs "huge" analysis to be analyzed [36] (Figure 4.1).

A photo pre-processing module (IPM), CDDM, and CYPM make up the SCYP model. As shown in Figure 3, the IPM captures leaf photos from net crawl, drone

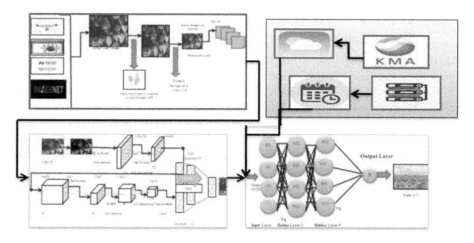

FIGURE 4.1 SCYP structure.

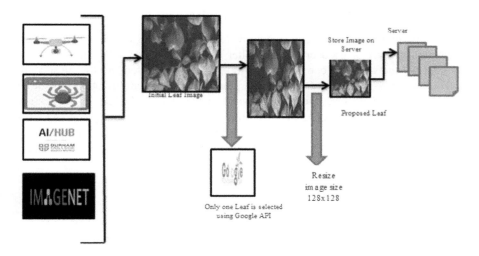

FIGURE 4.2 Image Processing Module (IPM).

image, AI server, and a series of photograph, then enhances the snippets using image compression and the Google Vision API. The normalized pictures are held on a computer server where a CNN version is created by the CDDM, and then learned from IPM-consolidated photographs and used for crop illness detection. The CYPM forecasts crop yields using CDDM-recognized pests, cutting-edge, temperature, and entirely ANN-based crop popularity data methodologies. In addition to precipitation and sunlight, and crop credibility statistics, including crop call, harvest date, and water pH, the CYPM receives meteorological data from the weather report from the farm server [37] (Figure 4.2).

4.4.2 Detection and Classification of Crop Diseases Using CNN

The CDDM that can diagnose crop diseases with the help of producing CNN fashions will be discussed in this chapter. The CDDM uses 30 epochs in the produced CNN models with the leaf snapshots pre-processed using IPM. Within each plant, the CDDM creates CNN architecture. A 128×128 picture is split into three (R, G, and B) to form a $128 \times 128 \times 3$ input. With the exception of the multitude of infections each crop suffers, there seems to be no collection of CDDM output nodes. The CDDM method's output price varies from zero to 1, and the most probable node is calculated as a crop ailment (Table 4.2).

Through the three convolutions and three max-pooling methods, and along with the complete CNN processes with the three layers, the CDDM develops a CNN model. The mechanisms of convolution and max-pooling wring out its diagnostic disease from the leaves, and the NN determines the last and final end result, the disease name. The process of convolution scans pictures to wring their features out, and to ensure development of convolutionary phenomenon charts [89, 90] (Figure 4.3).

4.4.3 Crop Yield Prediction Module (CYPM) Using ANN

Within the convolution process, a picture filter works with the help of either an orthogonal matrix of $n \times n$ or a Stride filter transfer unit. A 4×4 filter is used by the CDDM: the very first convolution method employs 20 filters, 40 for the second, and 60 for the last. In each of these phases, the first one is used throughout the three convolution mechanisms. The algorithms for convolution calculate the utilization of its distinctive feature, rectified linear unit (ReLu) of both a photograph and a filter from the coevolutionary function map. The CDDM deploys a 2×2 max-pooling methodology to optimize the convolutional characteristic map. In the same phase

TABLE 4.2
Compare Different of Classification Approaches

Classify Approach	Crop	Accuracy Rate in %
CNN Classifier	14 crops [38]	99.35%
CNN Classifier	Soybean [39]	99.32%
SVM Classifier	Citrus [40]	95% is actual level of service
	Grape [41]	88.89%
	Oil Palm [42]	With chimaera, 97% and 95% for Anthracnose disease.
	Soybean [43]	90%
KNN Classifier	Cotton [44]	82.5%
	Cherry, mango, pear, apple and grapevine [45]	96.3%
ANN Classifier	Groundnut [46]	97.41%
	Pomegranate [47]	About 90%

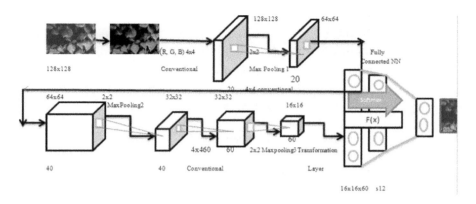

FIGURE 4.3 Crop Disease Diagnosis Module (CDDM).

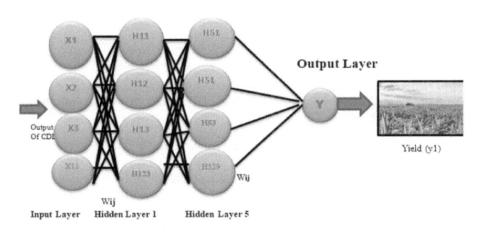

FIGURE 4.4 Crop Yield Prediction Module (CYPM).

as the current multilayer neural community, all nodes within the completely related neural community are computed. The CDDM uses SoftMax to assess the use of the SoftMax loss property as an enabling characteristic of the CNN or learn from it. A spectrum between 0 and 1 results in SoftMax. The CDDM helps clients to designate their underlying condition on the nodes greater than 0.75 and retain outlining quantities inside the output desk. The CDDM output table values, except 0.75 to 0, get from a farming system a pestilence summoned with an output desk adequate to or more than 0.75, and are contained inside this output file. The output sheet is transferred by the CDDMM into the CYPM (Figure 4.4).

Utilizing statistics like crop sickness, cutting-edge environment, and crop status, the CYPM module was used to forecast crop yield. The CYPM uses the outputs of the CDDM (including ailment names and pathogens), the name of the crop, the final harvest date, temperature, moisture, rainfall, sunlight, atmospheric tension, condensation, and pH value of water, including surface surrounding pH of soil. Table 4.3

TABLE 4.3
Input CYPM Nodes

Input	Explanation	Source
x_1	Disease 1	CDDM
x_2	Disease 1'Infectious	
x_3	Disease 2	
x_4	Disease 2'Infectious	
...	......	......
x_{2n+1}	Normal	
x_{2n+2}	Perception	
x_{2n+3}	Moisture	
......	......	......
x_{2n+10}	Date remaining to harvest	
x_{2n+11}	pH of Water	Server
x_{2n+12}	Quality of Water	
x_{2n+13}	pH of Soil	

shows that perhaps the scope of input layer nodes in the CYPM is $2n + 13$, where n shows the lot of illnesses that crops could have. Equation 4.1 shows all entry nodes within the CYPM:

$$X = \{X_1 + X_2 + X_3 + X_4 + X_5 + \cdots X_{2n+3}\} \tag{4.1}$$

The performance of the CYPM indicates the estimated production of crops; such that the number of available output access points and within CYPM should be 1, and along with this, five secret layers are used by the CYPM. Each node has 20 nodes, as given in Equation (4.2). The hidden level is represented by the superscript (m), while the node number is represented by the subscript (1, 2,..., and so on).

$$h^m = \{h_1^m, h_2^m, h_3^m, \ldots, h_{20}^m\} \tag{4.2}$$

In CYPM, every node is related to or associated with each other and the strain among the nodes is represented in Equation (4.3), with $w_{1,1}^1$ this equation showing the weight among the nodes of input layer.

$$W = \{w_{1,1}^1, w_{1,2}^1, w_{1,3}^1, \ldots w_{20}^6\} \tag{4.3}$$

The loss feature of CYPM is show in Equation (4.4):

$$E = \frac{1}{n}\sum_{k}^{n}\left(y_k - d_k^2\right) = \left(y - d\right)^2 \tag{4.4}$$

The activation function of ReLu is shown in Equation (4.5):

$$F(x) = \max(0, x) \tag{4.5}$$

The speed of computation is very high, but the result of the nodes is coming out negative, because in the above equations, E and ReLu have low accuracy. Both quantities positive (input and output) of the CYPM are shown in Equations (4.4) and (4.5) and guarantee velocity without deterioration.

$$\delta_y = (d - y) \tag{4.6}$$

The CYPM measures the δ_y fallacy signal using the expense of the output node for propagation. The δ_y fallacy sign is obtained in Equation (4.6) where learning data is d and the CYPM output price is y. If δ_y is determined, the CYPM computes the $\delta_{h_i^5}$ and fallacy signal of the last hidden layer by using δ_y is determined.

$$\delta_{h_i^c} = \sum_{k=1}^{28} \delta h_k^{c+1} w_{i,k}^{c+1} \tag{4.7}$$

$$\delta_{h_i^5} = \delta_y w_i^6 \tag{4.8}$$

In Equation (4.8), the fallacy sign $\delta_{h_i^5}$ is occupied by the left of hidden layer, and the main to the fourth hidden layer's fallacy sign δh_c^i in Equation (4.7). If all layers are determined for fallacy warnings, the weight between nodes is using the CPYM model. In Equation (4.11), between the hidden layers in Equation (4.9), and between the hidden layer and the entry layer in Equation (4.9), the weight between the remaining hidden layer and the output layer is modified (4.10). Within two and five, the k of Equation (4.9) is an integer.

$$w_{i,j}^k = w_{i,j}^k + \alpha h_i^{k-1} \delta_{h_j^k} \tag{4.9}$$

$$w_{i,j}^1 = w_{i,j}^1 + \alpha x_i \delta_{h_j^i} \tag{4.10}$$

$$w_i^6 = w_i^6 + \alpha h_i^5 \delta_y \tag{4.11}$$

4.4.4 Advantages of SCYP

Subsequent benefits are given to the SCYP over available research techniques. Firstly, despite previous findings which always treated successfully only one bacterial infection, SCYP could diagnose the diffusion of diseases that are more likely to arise with greater precision. Secondly, the SCYP can veraciously predict crop yields by using ailment and ambient interior data on vegetation. Throughout that whole section, harvest facts were common to diagnose crop sicknesses and predict vegetation beyond time, but plants should be recognized and yields need to be predicted on real farms in destiny research. The techniques of embellishing operating time at the same time ensuring the precision of the CNN used must also be researched in CDDM.

4.5   ROLE OF AI- AND ML-BASED ALGORITHMS FOR CROP DISEASE DETECTION

4.5.1   CNN-Based Analysis of Crop Disease Detection

With the assistance of Yann LeCun, a postdoctoral computing researcher, CNN, also known as ConvNets, was first published in the 1980s (Figure 4.5). At the work completed by way of Kunihiko Fukushima, a Japanese scientist who had formulated the acknowledgment more than a year earlier, LeCun created a sincerely primary image credibility neural community [48]. A much-regarded visual recognition and diagnostic set of rules are miles away. CNNs have become one of the most valuable creations that express the deep revival of the NN in laptop vision, which could be a fixed device to learn [49]. Compounded with creative and prescient computers, CNN is capable of not only conducting daunting operations from diagnosing pictures to solving astronomy clinical problems and designing self-riding vehicles [50]. The development of CNNs asked the qualities to hunt down rich mid-degree image showiness as to low-degree hand photo designed options are using a one-of-a-kind image form method [51]. Within the discipline of picture class, CNN has obtained astonishing results. A special way of schooling and the system used to promote the application of a fast and smooth device is depicted in exercise [52]. CNN's convolution layers can be visible as equivalent filters that could be run from the facts immediately. Thus, CNN creates a hierarchy of observable representations that can be tailored for a specific challenge. One of the predominant strengths of CNNs is its generalization ability, i.e., the ability to document methods never previously determined [53]. Initially, all the pictures are not inherited from the farmer. The photographs are obtained from either the farmworker through the use of the android functionality or a website created specifically for the activity of the farmer. An appropriate image of a leaf of the crops is captured. Then, the image is cropped to the best layout, and then it is likely to be submitted to the server on which CNN is used to enforce a set of rules [54]. Various diseases of plant leaves are detected in cotton, sugarcane, wheat, grape, etc. Mobile Net algorithm is used to train the data. Collaborators work with an accuracy of 97.33% [55].

4.5.2   Role of ANN in Analysis of Crop Disease Detection

In recent years, the creative and prescient assembly of the world's meals requirements for the increasing population of the earth is becoming more relevant. In the

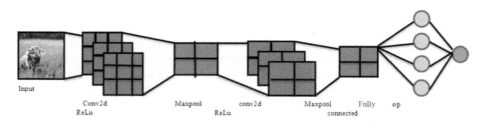

FIGURE 4.5   Basic CNN Module.

agricultural sector, crop models and decision equipment are increasingly used to increase production quality. Recently, the combination of advanced age and farming to embellish the crop yield meeting is becoming more interesting. Due to rapid progress of the new better generation, it can be predicted that crop fashions and predictive tools will become an integral component of precision agriculture [56]. Around 40 years ago, the ANN was pioneered and nowadays there is a first-rate interest in neural networks, considering a man-made network stocks a variety of biological components of the body and behavior [57]. The ANN method, which is a parallel system, is based on an organic neural technique of the human mind that will not solve complex problems in which it attempts to mimic mathematical models. Currently, due to its ability to expect, forecast, and class in biology fields, ANN has grown into a popular approach for most writers. Although it takes longer to create the regression version, instead of regression fashions, an ANN version can generate extra, unchanging yet accurate crop yield prediction [58]. When the ANN was used to expect the AUDPC in other studies, the regular correlation became 0.94 with estimates and 0.96 with three views, among the ANN's predicted and perceived values in six reviews [59]. The enforcement of five ANN models—generalized feed forward, multilayer perceptron (MLP), Jordan/Elman, principal component evaluation, and radial foundation feature—with wonderful learning algorithms transfer features, hidden layers, and neurons in each layer, along with multilinear regression models. Using statistical first-rate parameters, including coefficient of determination, root method and rectangular blunders, the management of the models in predicting seed yield changed to resolve and indicate absolute blunders. According to sensitivity analysis, the most influential attribute in the MLP/ANN version is the plant-compatible amount of pills (NCP), which is the seed production [60].

4.6 CASE STUDIES

4.6.1 Disease Detection on the Leaves of Tomato Using Deep Learning

The tomato (*Solanum lycopersicum*) is a fruit from the long circle of autochthonous relatives to the south of us in the nightshade. In addition, they have an excellent supply of vitamin C, potassium, folate, and k-diet[61]. Tomatoes are now the fourth most common fresh vegetable on the market next to potatoes, lettuce, and onions. After the discovery of the tomato's origins from America, the Spaniards delivered it in Europe. Initially, due to the fruit, it was delivered as a non-useful species. Its use as an agricultural plant, in the 1800s, has become comparatively recent. Its subculture spread within the nineteenth century, but it had not become an integral culture until 1900s. China, America, Turkey, Egypt, and Italy are its main foreign generating locations [62]. These are the statics listed in India until 2020 [63] (Figure 4.6).

4.6.2 Apple Leaf Real-Time Detection Using Deep Learning on CNN

Apples are amongst the most popular fruits in the world. They stretch over the apple tree (*Malus domestica*) mainly from central Asia. You've heard it a zillion

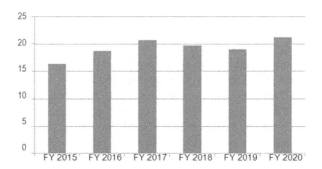

FIGURE 4.6   Statics till 2020 for tomato production.

instances: "An apple a day keeps the doctor away." There's more of a reality in its availability than you might assume. 89,565,973 tons of apples are produced annually worldwide. With an annual production volume of 44,448,575 tons, China is the world's largest apple producer. With 649,323 tons per year of production, the United States of America is the second one. Collectively, China and America generate about 50% of the world's output. India ranks at number 5 with its production of apples at 2,872,000 [64].

4.6.3 IDENTIFICATION OF RICE DISEASES USING DEEP CNN

Rice (*Oryza sativa*) is a grass plant (in the Poaceae family) cultivated widely for its starchy cereal grains ideal for eating. Approximately, half of the earth's population, including all of East and Southeast Asia, certainly depends on rice as a staple food; 95% of the rice crop in the sector is eaten by people [65]. Rice, with around 761.5 million tons (1,000 kg) produced in 2018, is the rural object with the third highest global production worldwide. Rice is grown in 120 locations, but China (approximately 214 million tons) and India (approximately 173 million tons) contributes nearly half of global production. Southeast Asia is home to 9 of the top 10 and 13 of the top 20 rice-producing nations in the world [66]. The traditional approach to distinguish between rice diseases involves the discovery and expertise of several specialists. We've researched the credibility and system study of automated rice sickness analysis assisted by the pattern. It includes the use of pattern recognition techniques [67], vector support machines [68], digital image processing techniques [69], and computer vision [70]. A fully specific technique for identifying rice sicknesses, backed by deep conventional neural networks, was suggested. An elevated class ratio may be obtained by the most up-to date CNN-dependent edition. First, CNN model may also compose of ten or more layers, each layer containing heaps of neurons. The second is that large-scale database, such as image net ILSVRC, contains many educational examples that remain to be awaited by well-prepared, deep, acquiring knowledge of the collection of laws. This version of CNN will enhance the speed of convergence, while learning the characteristics in CNNs and achieving more prominence accuracy than the traditional version.

The maximum contribution of these studies is especially twofold:

- A CNN is used first to resolve the question of identity of rice disease.
- Experiment results display that the CNN approach cannot best improve the convergence pace; however, it can additionally obtain higher reputation accuracy than other different models, such as:
  - Support vector machine (SVM)
  - Standard BP algorithm
  - Particle swarm optimization (PSO) [71].

4.6.4 IMAGE-BASED POTATO TUBER DISEASE DETECTION

Every year potato (*Solanum tuberosum*) grows inside the nightshade familgroup (Solanaceae), and is planted for its starchy protection. The plant was a vital crop in Ireland in the 17th century, and it had become a widely cultivated crop in the 18th century throughout continental Europe, primarily in Germany, and in the west of England [72]. Mass potato production took place in India in the period from 2015 to 2019, with a target date of 2020 [73]. Potato tubers can be infected with a variety of seeds and soil-borne disease that impacted tuber first rate and production, popular scrubs responsibility. Both of these diseases concentrate on tuber imperfections, seeds tubers are frequently damaged like typical retail tubers for processing and trading [74]. This example takes a look at developments in computer vision and visible perception of anchorage to be used in categorizing a few diseases in potato tubers (Figure 4.7).

For the image style task, the algorithm chosen was transformed into a deep CNN. The rudimentary option of the project for this muddle turned into the CNN model developed by the Oxford College Visual Geometry Group (VGG), called CNN-F because of its fasting schooling timings [75]. Due to the relatively restricted data set, multiple extra drop outs layers were added to the VGG architecture to address problems of over-becoming. For statistical augmentation, two techniques were used [76]. In this project, in order to classify physiological reinforcement in four ailment classes of disease-ridden potato tubers, the applicability of a CNN and a visually healthy beauty

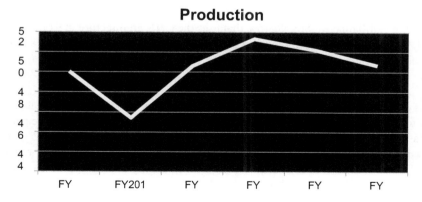

FIGURE 4.7 Potato production.

has been scrutinized. The findings show that the appropriate class of fully coached CNN fashions varies from 83% for the version taught on the least amount of statistics to 96% for the version taught on 90% of the statistics. It's far enough to apply 493 images from our data collection to achieve classification costs greater than 90%. The results on texture features recommend that even a scaling factor or object tracking category composed of faster R-CNNN is paired with a set of protocols [77] that will promote detection of disease in images of whole tubers. Continuing studies seek to create a shaping algorithm with an accelerated amount of instructions for the disorder. Statistics can be obtained without problems since there are no data acquisition constraints [76,87,88]..

4.6.5 AUTOMATIC DETECTION OF LEAF SPOT DISEASE IN SUGAR BEET USING DL

Sugar beet might have been a massive quadrennial farming crop used to produce sugar. Approximately 30% of sugar produced worldwide comes from sugar beet [78]. Sugar beet production of 279,396,160 tons corresponds to a year's worth of production. The Russian Federation is the largest sugar beet producer with an annual production volume of 51,366,830 tons. With 33,794,906 tons of production per year, France is at the second place. India does not produce beet sugar [79]. Cercospora leaf spot, Fusarium, Rhizomania, and Rhizoctonia crown and plant disease are some of the standard diseases affecting sugar beet. The sickness of the leaf spot begins on the outer sides of leaves, then extends into the inner leaves, and dries out sugar beet leaves perennially [80]. Disease-related yield losses range from 10% to 50%, simply based on the nature of disease [81]. Consequently, signs of the condition must be identified at a proper time, and relevant steps must be taken immediately to avoid further development or worsening of the disease. Such an efficient and comprehensive diagnostic would fully eliminate yield degradation in sugar beet research domains [82]. In the studies, it is suggested that deep knowledge of the technique provides greater accuracy and higher overall performance than the normal image-processing strategies earlier used. The previous findings depict, 92.88% accuracy in 3,700 images [83], 96.3% accuracy in 2589 images [84], 93.88% accuracy, using AlexNet-based framework, and 95% and 47% accuracy, using GoogLeNet-based model with 54,306 images[85], 95.54% accuracy using the CNN model in 9,000 images [86], 96.46% that uses the CNN architecture in 9,000 images [86], 96.46% accuracy using the GoogLeNet-based model in 54 306 images [85].

4.7 CONCLUSION AND FUTURE SCOPE

This chapter presents the SCYP version for crop sickness, which employ DL technology to diagnose crop disease through crop leaf image and forecasts the whole harvest for a family farm based on end product diagnosis and treatment, as well as meteorological parameters. Four tests have been evaluated in order to verify their reliability of the SCYP. The CDDM's CNN precision is roughly 3.5% significantly greater than the one of R-CNN and 5.4% higher than that of YOLO in the first test. CNN's running time is approximately 195.5 seconds lengthier than the one of R-CNN and 160 seconds longer than among YOLO in the 2D research. The CDDM uses CNN, which would be more appropriate for tracking illnesses, especially given the fact that it would be sluggish than R-CNN and YOLO. The ReLu used by the CYPM (Crop Yield Prediction Version) had an accuracy approximately 1% greater than the sigmoid feature and approximately 10% greater than the Phase feature in the third test. In the fourth experiment, the CYPM predicted that by using multiple diseases, yields were around 34% more accurate than using people when they are not available. Hence, the SCYP can more precisely predict farm yields than conventional approaches.

REFERENCES

1. Kambale, G., & Bilgi, N. (2017). "A survey paper on crop disease identification and classification using pattern recognition and digital image processing techniques." IOSR Journal of Computer Engineering, 14–17. vol 4
2. "Agriculture and Allied Industries" [online]. Available at https://www.ibef.org/archives/industry/agriculture-reports/indian-agriculture- industry-analysis-january-2020
3. "Agriculture in India: Information about Indian Agriculture & Its Importance" [online]. Available at https://www.ibef.org/industry/agriculture-india.aspx
4. "India GDP from Agriculture" [online]. Available at https://tradingeconomics.com/india/gdp-from-agriculture
5. "India Needs Farm Revolution to Attain 9–10% GDP Growth: Amitabh Kant 2020" [online]. Available at https://economictimes.indiatimes.com/news/economy/agriculture/india-needs-farm- revolution-to-attain-9-10-gdp-growth-amitabh-kant/articleshow/68473771.cms?from=mdr
6. "Agriculture Sector to Grow at 3% in 2020–21 Despite Covid-19 Lockdown: Modi Govt" [online]. Available at https://theprint.in/india/agriculture-sector-to-grow-at-3-in-2020-21-despite-covid-19-lockdown-modi-govt/411458/
7. Han, L., Haleem, M. S., & Taylor, M.. (2015). A novel computer vision-based approach to automatic detection and severity assessment of crop diseases. *2015 Science and Information Conference (SAI)*. IEEE.
8. Aurangzeb, K., Akmal, F., Khan, M. A., Sharif, M., & Javed, M. Y. (2020, March). Advanced machine learning algorithm based system for crops leaf diseases recognition. In *2020 6th Conference on Data Science and Machine Learning Applications (CDMA)* (pp. 146–151). IEEE.
9. Kaur, R., & Kang, S. S. (2015, October). An enhancement in classifier support vector machine to improve plant disease detection. In *2015 IEEE 3rd International Conference on MOOCs, Innovation and Technology in Education (MITE)* (pp. 135–140). IEEE.
10. Dhanachandra, N., Manglem, K., & Chanu, Y. J. (2015). Image segmentation using K-means clustering algorithm and subtractive clustering algorithm. *Procedia Computer Science*, *54*, 764–771.
11. Chouhan, S. S., Kaul, A., Singh, U. P., & Jain, S. (2018). Bacterial foraging optimization based radial basis function neural network (BRBFNN) for identification and classification of plant leaf diseases: An automatic approach towards plant pathology. *IEEE Access*, *6*, 8852–8863.
12. Wang, J., Yang, Y., Mao, J., Huang, Z., Huang, C., & Xu, W. (2016). Cnn-rnn: A unified framework for multi-label image classification. In *Proceedings of the IEEE conference on computer vision and pattern recognition* (pp. 2285–2294).
13. *Plant Village*, Available at https://plantvillage.psu.edu/, Accessed: 02/12/2020.
14. Al Bashish, D., Braik, M., & Bani Ahmad, S. (2011). Detection and classification of leaf disease using K-mean based segmentation and neural network based classification. *International Technology Journal*, 267–275. ISSN 1812-5638.

15. Ji, S., Zhang, C., Xu, A., Shi, Y., & Duan, Y. (2018). 3D convolutional neural networks for crop classification with multi-temporal remote sensing images. *Remote Sensing, 10*(1), 75–84.

16. Gupta, D., & Ahlawat, A. (2017). Usability feature selection via MBBAT: A novel approach. *Journal of Computational Science, 23*, 195–203.

17. Singh, V., & Misra, A. K. (2017). Detection of plant leaf diseases using image segmentation and soft computing techniques. *Information Processing in Agriculture, 4*(1), 41–49.

18. Pantazi, X. E., Moshou, D., & Tamouridou, A. A. (2019). Automated leaf disease detection in different crop species through image features analysis and One Class Classifiers. *Computers and Electronics in Agriculture, 156*, 96–104.

19. Wen, J., Shi, Y., Zhou, X., & Xue, Y. (2020). Crop disease classification on inadequate low-resolution target images. *Sensors, 20*(16), 4601.

20. Kadir, A. (2014). A model of plant identification system using GLCM, lacunarity and shen features. . *5*(2). arXiv preprint arXiv:1410.0969.

21. Kamal, K. C., Yin, Z., Li, B., Ma, B., & Wu, M. (2019, September). Transfer learning for fine-grained crop disease classification based on leaf images. In 2019 10th Workshop on Hyperspectral Imaging and Signal Processing: Evolution in Remote Sensing (WHISPERS) (pp. 1–5). IEEE.

22. Morbekar, A., Parihar, A., & Jadhav, R. (2020, June). Crop Disease Detection Using YOLO. In 2020 International Conference for Emerging Technology (INCET) (pp. 1–5). IEEE.

23. Raikar, K., Gawade, S., & Turkar, V. (2017, October). Usability improvement with crop disease management as a service. In 2017 International Conference on Recent Innovations in Signal *Processing* and Embedded Systems (RISE) (pp. 577–582). IEEE.

24. Yun, S., Xianfeng, W., Shanwen, Z., & Chuanlei, Z. (2015, August 25). PNN based crop disease recognition with leaf image features and meteorological data. *International Journal of Agricultural and Biological Engineering, 8*(4), 60–68.

25. Kulkarni, A. H., & Patil, A. (2012). Applying image processing technique to detect plant diseases. *International Journal of Modern Engineering Research, 2*(5), 3661–3664.

26. Pujari, J. D., Yakkundimath, R., & Byadgi, A. S. (2014, December). Identification and classification of fungal disease affected on agriculture/horticulture crops using image processing techniques. In *2014 IEEE International Conference on Computational Intelligence and Computing Research* (pp. 1–4). IEEE.

27. Rastogi, A., Arora, R., & Sharma, S. (2015, February). Leaf disease detection and grading using computer vision technology & fuzzy logic. In *2015 2nd international conference on signal processing and integrated networks (SPIN)* (pp. 500–505). IEEE.

28. Singh, A. K., Rubiya, A., & Raja, B. S. (2015). Classification of rice disease using digital image processing and SVM classifier. *International Journal of Electrical and Electronics Engineers, 7*(1), 294–299.

29. Sabrol, H., & Satish, K. (2016, April). Tomato plant disease classification in digital images using classification tree. In *2016 International Conference on Communication and Signal Processing (ICCSP)* (pp. 1242–1246). IEEE.

30. Sharma, R., Singh, A., Dutta, M. K., Riha, K., & Kriz, P. (2017, July). Image processing based automated identification of late blight disease from leaf images of potato crops. In *2017 40th International Conference on Telecommunications and Signal Processing (TSP)* (pp. 758–762). IEEE.

31. Periasamy, S., & Shanmugam, R. S. (2017). Multispectral and microwave remote sensing models to survey soil moisture and salinity. *Land Degradation & Development, 28*(4), 1412–1425.

32. Park, H., JeeSook, E., & Kim, S. H. (2018, August). Crops disease diagnosing using image-based deep learning mechanism. In *2018 International Conference on Computing and Network Communications (CoCoNet)* (pp. 23–26). IEEE.

33. "Farm Progress" [online]. Available at https://www.farmprogress.com/crop-disease/6-major-issues-threaten-future-crop-disease-management

34. "Pest and Disease Cause Worldwide Damage of crop" [online]. Available at https://californiaagtoday.com/pests-diseases-cause-worldwide-damage-crops/

35. "Use case: Precision Agriculture, the Internet of Things, and Big Data Management" [online]. Available at https://helioswire.com/case-study-precision-agriculture-the-internet-of-things-and-big-data-management/

36. "Precision Ag & Big Data Learning" [online]. Available at https://www.precisionag.com/systems-management/data/precision-ag-big-data-learning/

37. Lee, S., Jeong, Y., Son, S., & Lee, B. (2019). A self-predictable crop yield platform (SCYP) based on crop diseases using deep learning. *Sustainability, 11*(13), 3637

38. Mohanty, S. P., Hughes, D. P., & Salathé, M. (2016). Using deep learning for image-based plant disease detection. *Frontiers in Plant Science, 7,* 1419.

39. Shruthi, U., Nagaveni, V., & Raghavendra, B. K. (2019, March). A review on machine learning classification techniques for plant disease detection. In *2019 5th International Conference on Advanced Computing & Communication Systems (ICACCS)* (pp. 281–284). IEEE.

40. Gavhale, K. R., Gawande, U., & Hajari, K. O. (2014, April). Unhealthy region of citrus leaf detection using image processing techniques. In *International Conference for Convergence for Technology-2014* (pp. 1–6). IEEE.

41. Padol, P. B., & Yadav, A. A. (2016, June). SVM classifier based grape leaf disease detection. In *2016 Conference on Advances in Signal Processing (CASP)* (pp. 175–179). IEEE.

42. Masazhar, A. N. I., & Kamal, M. M. (2017, November). Digital image processing technique for palm oil leaf disease detection using multiclass SVM classifier. In *2017 IEEE 4th International Conference on Smart Instrumentation, Measurement and Application (ICSIMA)* (pp. 1–6). IEEE.

43. Kaur, S., Pandey, S., & Goel, S. (2018). Semi-automatic leaf disease detection and classification system for soybean culture. *Journal on IET Image Processing, 12*(6), 1038–1048.

44. Parikh, A., Raval, M. S., Parmar, C., & Chaudhary, S. (2016, October). Disease detection and severity estimation in cotton plant from unconstrained images. In *2016 IEEE International Conference on Data Science and Advanced Analytics (DSAA)* (pp. 594–601). IEEE.

45. Sladojevic, S., Arsenovic, M., Anderla, A., Culibrk, D., & Stefa-Novic, D. (2016). Deep neural networks based recognition of plant diseases by leaf image classification. *Computational Intelligence and Neuroscience*, Article ID: 3289801.

46. Ramakrishnan, M. (2015, April). Groundnut leaf disease detection and classification by using back probagation algorithm. In *2015 International Conference on Communications and Signal Processing (ICCSP)* (pp. 0964–0968). IEEE.

47. Dhakate, M., & Ingole, A. B. (2015, December). Diagnosis of pomegranate plant diseases using neural network. In *2015 Fifth National Conference on Computer Vision, Pattern Recognition, Image Processing and Graphics (NCVPRIPG)* (pp. 1–4). IEEE.

48. "What Are Convolutional Neural Networks (CNN)" [online]. Available at https://bdtechtalks.com/2020/01/06/convolutional-neural-networks-cnn-convnets/

49. "Convolutional Neural Network" [online]. Available at https://docs.paperspace.com/machine-learning/wiki/convolutional-neural-network-cnn

50. "Introduction to Convolutional Neural Network (CNN) Using Tensor flow" [online]. Available at https://towardsdatascience.com/introduction-to-convolutional-neural-network-cnn-de73f69c5b83

51. Oquab, M., Bottou, L., Laptev, I., & Sivic, J. (2014, June). Learning and transferring mid-level image representations using convolutional neural networks. In *Proceedings of the 27th IEEE Conference on Computer Vision and Pattern Recognition (CVPR '14)* (pp. 1717–1724). IEEE, Columbus, OH.

52. Sladojevic, S., Arsenovic, M., Anderla, A., Culibrk, D., & Stefanovic, D. (2016). Deep neural networks based recognition of plant diseases by leaf image classification. Computational *Intelligence* and *Neuroscience*, 2016.

53. Boulent, J., Foucher, S., Théau, J., & St-Charles, P. L. (2019). Convolutional neural networks for the automatic identification of plant diseases. *Frontiers in* Plant Science, *10*, 941.

54. Neural Networks: H.A. Rowley, Department of Computer Science, Carnegie Mellon University, Pittsburgh, PA.

55. Machha, S., Jadhav, N., Kasar, H., & Chandak, S. (2020). Crop leaf disease diagnosis using convolutional neural network. International Journal of Trend in Scientific Research and Development (IJTSRD), 4. 1056–1058

56. Khairunniza-Bejo, S., Mustaffha, S., & Wan Ishak Wan Ismail. (2014). Application of artificial neural network in predicting crop yield: A review. *Journal of Food Science and Engineering, 4*(1), 1.

57. Marchant, J. A., & Onyango, C. M. (2002). Comparison of Bayesian classifier with multilayer feed-forward neural network using example of plant/weed/soil discrimination. *Computers and Electronics in Agriculture, 39*, 3–22.

58. Kaul, M., Hill, R. L., & Walthall, C. (2005). Artificial neural network for corn and soybean prediction. *Agricultural System 85*, 1–18.

59. Alves, D. P., Tomaz, R. S., Laurindo, B. S., Laurindo, R. D. S., Silva, F. F. E., Cruz, C. D., Nick, C., & da Silva, D. J. H. (2017). Artificial neural network for prediction of the area under the disease progress curve of tomato late blight. *Scientia Agricola, 74*, 51–59.

60. Abdipour, M., Younessi-Hmazekhanlu, M., RezaRamazani, S. H., & Omidi, A. H. (2019). Artificial neural networks and multiple linear regression as potential methods for modeling seed yield of sa_ower (Carthamus tinctorius L.). *Industrial Crops and Products, 127*, 185–194.

61. "Tomatoes 101: Nutrition Facts and Health Benefits" [online]. Available at https://www.healthline.com/nutrition/foods/tomatoes

62. "Fruit & Vegetable" [online]. Available at https://www.frutas-hortalizas.com/Vegetables/About-Tomato.html

63. "Production Volume of Tomatoes across India from Financial Year 2015 to 2019, with an Estimate for 2020" [online]. Available at https://www.statista.com/statistics/1039712/india-production-volume-of-tomatoes/

64. "World Leading Apple Producing Countries" [online]. Available at https://www.atlasbig.com/en-in/countries-by-apple-production

65. "Rice" [online]. Available at https://www.britannica.com/plant/rice

66. "Rice Production by the Country in 2020" [online]. Available at https://worldpopulationreview.com/country-rankings/rice-production-by-country

67. Phadikar, S., & Sil, J. (2008, December). Rice disease identification using pattern recognition techniques. In *2008 11th* International Conference on Computer and Information Technology (pp. 420–423). IEEE.

68. Jian, Z., & Wei, Z. (2010, March). Support vector machine for recognition of cucumber leaf diseases. In *2010 2nd International Conference* on *Advanced Computer Control* (Vol. 5, pp. 264–266). IEEE.

69. Barbedo, J. G. A. (2013). Digital image processing techniques for detecting, quantifying and classifying plant diseases. *SpringerPlus, 2*(1), 660–672.

70. Asfarian, A., Herdiyeni, Y., Rauf, A., & Mutaqin, K. H. (2014). A computer vision for rice disease identification to support integrated pest management. *Crop Protection, 61,* 103–104.

71. Lu, Y., Yi, S., Zeng, N., Liu, Y., & Zhang, Y. (2017). Identification of rice diseases using deep convolutional neural networks. *Neurocomputing, 267,* 378–384.

72. "Potato" [online]. Available at https://www.britannica.com/plant/potato

73. "Production Volume of Potato across India from Financial Year 2015 to 2019, with an Estimate for 2020"[online]. Available at https://www.statista.com/statistics/1038959/india-production-of-potato/

74. Mattupalli, C., Genger, R. K., & Charkowski, A. O. (2013). Evaluating incidence of *Helminthosporium solani* and *Colletotrichum coccodesona* symptomatic organic potatoes and screening potato lines for resistance to silver scurf. *American Journal of Potato Research, 90,* 369–377

75. Chatfield, K., Simonyan, K., Vedaldi, A., & Zisserman, A. (2014). Return of the devil in the details: Delving deep into convolutional nets. arXiv1405.3531.

76. Oppenheim, D., Shani, G., Erlich, O., & Tsror, L. (2019). Using deep learning for image-based potato tuber disease detection. *Phytopathology, 109*(6), 1083–1087.

77. Ren, S., He, K., Girshick, R., & Sun, J. (2015). Faster R-CNN: Towards real-time object detection with region proposal networks. In C. Cortes, N. D. Lawrence, D. D. Lee, M. Sugiyama, & R. Garnett (Eds.), *Advances in Neural Information Processing Systems* 28 (pp. 91–99). Neural Information Processing Systems Foundation Inc., La Jolla, CA.

78. Geçit, H. H., Çiftçi, C. Y., Emeklier, H. Y., Ikincikarakaya, S., Adak, M. S., Kolsarıcı, Ö., Ekiz, H., Altınok, S., Sancak, C., Sevimay, C. S., & Kendir, H. (2011). Tarla Bitkileri, No: 1588, Düzeltilmiş İkinci Baskı, in: Ders Kitabı, vol. 540, Ankara Üniversitesi Ziraat Fakültesi Yayınları, Ankara.

79. "World Leading Sugar Beet Producing Countries" [online]. Available at https://www.atlasbig.com/en-in/countries-by-sugarbeet-production

80. Ozguven, M. M., & Adem, K. (2019). Automatic detection and classification of leaf spot disease in sugar beet using deep learning algorithms. *Physica A: Statistical Mechanics and Its Applications, 535*(2019), 122537.

81. Mohamed, F. R. Smith, K., & Larry, J. (2005). Evaluating fungicides for controlling cercospora leaf spot on sugar beet. *Crop Protection, 24,* 79–86.

82. Bock, C. H., Poole, G. H., Parker, P. E., & Gottwald, T. R. (2010). Plant disease severity estimated visually, by digital photography and image analysis and by hyperspectral imaging. *Critical Reviews in Plant Sciences, 29*(2), 59–107. doi:10.1080/07352681003617285

83. Amara, J., Bouaziz, B., & Algergawy, A. (2017). A deep learning-based approach for banana leaf diseases classification. *Datenbanksysteme für Business, Technologie und Web. BTW Workshop*, Stuttgart.

84. Sladojevic, S., Arsenovic, M., Anderla, A., Culibrk, D., & Stefanovic, D. (2016). Deep neural networks based recognition of plant diseases by leaf image classification. In: *Computational Intelligence and Neuroscience* (Vol. 2016, p. 11). Hindawi Publishing Corporation, Article ID: 3289801, doi:10.1155/2016/3289801.

85. Mohanty, S. P., Hughes, D. P., & Salathé, M. (2016). Using deep learning for image-based plant disease detection. *Frontiers in Plant Science, 7,* 1419.

86. Ashqar, B. A. M., & Abu-Naser, S. S. (2018). Image-based tomato leaves diseases detection using deep learning. *International Journal of Academic Engineering Research, 2*(12), 10–16.

87. Geetharamani, G., & Arun, P. J. (2019). Identification of plant leaf diseases using a nine-layer deep convolutional neural network. *Computers & Electrical Engineering, 76,* 323–338.

88. Gensheng, H., Haoyu, W., Yan, Z., & Mingzhu, W. (2019). A low shot learning method for tea leaf's disease identification. *Computers and Electronics in Agriculture, 163,* 104852.

89. Srivastava, S. (2018). Accurate human recognition by score level and feature level fusion using palm-phalanges print. *Arabian Journal for Science and Engineering, 43*(2), 543–554.

90. Chaudhary, G., Srivastava, S., & Bhardwaj, S. (2017). Feature extraction methods for speaker recognition: A review. *International Journal of Pattern Recognition and Artificial Intelligence, 31*(12), 1750041.

5 Architecture of Digital Twin for Network Forensic Analysis Using Nmap and Wireshark

Kapil Kumar and Manju Khari
Netaji Subhas University of Technology East
Campus, and Jawaharlal Nehru University

CONTENTS

5.1 INTRODUCTION

The persistency of data has crafted an opportunity to create a framework with potential that is similar to DT. The ideas of DT have arisen to style the corporeal entity and associated facts using reachable software so that customers can be reached over digital platforms. The nomenclatures consumed for parallel or overlying ideas comprise digital foil, simulated twin, simulated entity, and artifact. The DT contains three portions that include a corporeal entity, artifact, and the connection between the corporeal and artifact. The DT allows online commnique with the somatic like in equal commands. The steady development in information technology, schmoozing

DOI: 10.1201/9781003132868-5

and interaction provides us rewards and enhances our need for computing devices, and on the Internet. The organization has day-to-day operations related to finance and marketing on the Internet due to which network security is extremely important (Mandia, K., & Procise, C. 2004). It is a giant problem in the structural system. Owing to the reliance of corporations on web services and the easy availability of the Internet, it has become a central target of cybercrimes. Computer customers need appropriate security to ensure the safety and security of their prized facts from online muggers. Palmer (2001) found that numerous methods are available to protect network structure and to do transmission across the network, like the controller for preclusion—the system to detect the intrusion. It is a preventive system to detect intrusion. Nonetheless, currently, groups are engrossed to identify and trail up the intruders. Enhancement in cybercrimes stretches the natal novel division as scientific or forensic.

Cyber forensics is a method that is related to identifying, amassing, stowing, examining, and viewing the outcome as an indication of legitimation (Pilli, E. et al. 2010). The forensic of a complex lattice-like network is a method to seize, stock, and examine sachets that come via a network from an analytical point. The authors (Kao et al., 2018) found that network forensic is a discipline that pacts with catching, logging, and tempting transmitted data for noticing and examining it. Network forensic is the next step in network security. The protection framework behaves as a guard contrary to the bout from invaders although forensic schema forms in the context of network and gathers evidence related to veracity like a sign related to intrusion. The area of forensics-related to the network is the zone for utilizing scientific discipline in the network milieu to discover the origin of cybercrimes. The goal is to spot doubtful action like the data attained from the network and determine whether it is a current or upcoming threat (Kaushik, A. et al., 2010).

The zone of forensics is demarcated to "the use of methodically confirmed method to store, combine, classify, analyze, relate, examine, and document digital evidence from various, aggressive dispensation and convey digital cradles for the tenacity of easing or advancing the rebuilding of actions to be illegal, or serving to forestall illegal movements exposed to be troublesome to strategic activities". Network forensics is the discipline that contracts with catch, logging, and examination of network traffic for finding interruptions and examining them. Network forensics is fundamentally about observing, taking, and examining the network traffic, and inspecting the security policy abuses. Singh et al. (2020) discussed how forensic professionals screen the network unceasingly and record a copy of all the pertinent packets, liable upon the policy, in an agreed format for the upcoming study. These stowed sachets data are expanded and examined either by hand or by customizing numerous methods, to discover if there is a glitch in the network and if those glitches are about to happen.

If any suspicious intrusion is discovered, the kind of bout is dogged and the source of the bout is examined (Yasinsac, A., & Manzano, Y. 2010). Forensic authorities can ascribe the invaders by good screening, capturing, and examination of network traffic, and applying investigation tools. The key target is to determine the origin of an intruder and deliver an indication to classify the intruders. Muggers probe numerous connections through diverse protocols like HTTP, UDP, and FTP, FTP is a tactic to know whether the connection is available or closed; if a connection is exposed then the next mugger can continue for the bout (Mandia, K., & Procise, C. 2004).

Mugger utilizes denial-of-service (Dos) bout. Our target is to examine port scanning bouts and learn the origin of intruders. The problem statement indicates the malware could be discovered of the highest risk that can lead to leakage of information to private and imperative resources.

It is found that malware is a lively plugin that efforts to penetrate the device by breaking privacy. The malware attempts to violate the rule of security by using weak points of supports of security as the CIA is required to clutch the security (Chandran, R. 2004).

The authors proposed a framework to analyze the risk of attacks and their detection of the attacks using Nmap and Wireshark. The central intent is to improve upon the network forensic analysis method for detecting malware while investigating through Nmap-related tools. The objectives include designing a framework by using captured data, storing data, and analyzing the data to detect the malware. Analysis of the designed framework for the network by using Nmap, Wireshark tools, and detection accuracy of attacks needs to be upgraded through analysis.

Hence, the significant inclusion is to use of a Wireshark for investigating the data captured from the network and using optimal attributed to performing the analysis of assorted packets. The framework would be applicable in detecting intrusion through the network analysis and to detect the source of bogus information as it brings us to the target by using DDoS attacks and botnet.

The remaining segments of the chapter are systematized as shadows: The second segment presents a literature review; while the third segment proposes the effort. The fourth is about experiments and results. The fifth is about research limitations. The sixth is about the conclusion and future scope, and the final about references.

5.2 LITERATURE REVIEW

By the massive use of information technology in each sector, it is essential to uphold high-level safety to confirm, safe, and reliable interaction of information between many organizations and properties, Consequently information transferred over the internet and another network is under the threat of intrusions and abuses. In this section, numerous research papers are explored about malware and security.

Palmer (2001) proposed a framework for forensic discipline in the network milieu in the context of recognition, protection, assortment, inspection, investigation, performance, and result. In another paper, Reith et al. (2011) planned a framework on abstract digital forensic recognition, research and tactic pan, protection, assortment, inspection, analysis, performance, and returning sign.

Shu et al. (2017) described that fake news is today rolling out at a very fast rate and has become a complex problem in a small-time on diverse podia. Regularly these kinds of updates are of a stumpy class that contributes to false information. Due to broad rollout of fake news, it damagingly influences different groups. A tremendously fast dispersing fake news on social media or any other platforms can roll out buzzes, and revolts in humanity that don't have faiths. Fake news is a portion of news that is not properly explored or measured at any stage so provides distorted information or tricks its consumer through diverse methods of news delivery to multiple targets on networks.

The sources (Chandran, R. 2004) found that the sequence of extracting information from the linked machine is named network data mining. It comprises identification, collection, and analyzing the fetched data. The packet inspection of the grabbed torrent of traffic discourse about controlling the traffic is produced by a reliable cradle or was decided by the bots. Networking device comprises complex information, which is prized in a review of cybercrime situation. Every networking device is strong and long-lasting. There are phases to be straggled to inspect such rules like Firewalls, L3 switches, Intrusion Prevention Systems (IPS), etc. The tactics must be shaped to diminish and explore forensic glassy IoT devices. Because of renown efficiency, IoT is improving among both customers and interlopers. It is typically found that the Internet is not an inoffensive podium, which is full of hazardous risks, with prices in all ranges.

Prosise et al. (2003) projected a combined digital study course framework with stages as equipped, review, quest, gathering, procedure, reformation, and logging. These stages aim to safeguard that privacy and the foundation are capable of complete sustenance as per study when an event happens. Equipped stages reveal the eagerness of action structure, it has further stages such as review, rifle, assortment pool, and manage the data. Then there is a need to scrutinize the information and log the proof. In this paper (Casey and Palmer 2004), an investigation is proposed to study the framework with the appraisal phase to authenticate the event and conclusion stages about to look into whether they can revive the inspection or not. They are augmented with the shared stage.

Bagyalakshmi et al. (2018) proposed study to build a dynamic forensically complete course for WRI. This involves firm and consistent events for a canvasser to trail while trying to create this method as dynamic as possible by reducing social interaction.

Thus, it confirms the veracity and legitimacy of the information. To attain this, both gaining and investigation tools were established. A relative study of the advanced tool to prevailing tools exposed upgraded enactment in the context of phase and permissibility. Singh et al. (2020) emphasized delivering the set of prevalent databanks presented ranging from touch to neurological motions, to intellect image gathering, and so on. The amount and the classes of bouts obliquely interact with palmtop schemes that are rambled with an implication of security of information.

5.3   PROPOSED FRAMEWORK

The projected framework of the Network Forensic System has three sections encompassing an assortment and safeguarding section, the analysis section, and the view section are represented in Figure 5.1. The operation of the scheme is aimed at gathering the sachets via links, examine for the prized sachets, and to unveil evidence related to doubtful addresses.

The above figure shows the network forensic analysis (NFA) approach that includes information captured from the network, collected, and preserved using the Wireshark. Subsequently the data is analyzed by using Nmap to determine internet address details (IP details).

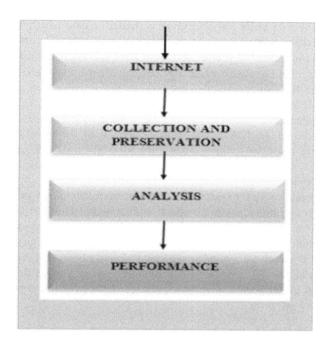

FIGURE 5.1 Framework of network forensic analysis.

It has numerous functions as an assortment and the key job of the assortment and safeguarding section are to grasp traffic from the links for analysis. This section amasses complete sachets that are conveyed via links of a network of the emcee concern system for passing them toward UNIX interrelated communique sequentially. The functioning of first in and first out is based on reciting of data that is occurring and rejecting. The sachets are caught from links through the Wireshark API, and the outcome is attained through Wireshark needs to send operation of first in and first out. Control function is aimed at receiving, and sending sachets written as "Wireshark -v try.txt". The setup is related to obtaining sachets. In this section sifting, and mere protocol of transmission control sachets will have fallouts in smaller storing area of memory as linked to collect sachets transferring via links.

Earlier utilized encoding for privacy in the contextual collection of network dashes to uphold the veracity. But currently, the section discusses, the manuscript that is transmuted and is keen on image and offers privacy, along with the usage of the dynamic scale of the waterline that delivers veracity, and measures out from scale. It is related to the waterline, and there is a need for a safety pattern to deliver validation.

The second stage is the analysis section that proceeds as input. It records and stores data in the host system. This section selects only TCP sachets for investigation. It measures bout through port screening method. TCP-SYN, TCP-ACK, TCP-FIN approaches are recognized by S, R, and F ensigned in sachets. A data stricture detail is well equipped with subsequent sections: apposite machine address reveals IP address that appeals for joining the host, port computation which includes an absolute number of ports for which link is invited, a date which means the date at which association invitation happens, initial time includes the instant when initial association invitation is directed, and complete-time which include instant when final association invitation is directed. The ensuing algorithm is used for the examination of sachets that are caught from network traffic.

Snag considers entire sachets that pounce on the host computer consuming the usefulness of Wireshark in a typescript file. Practice command Wireshark –t >try.txt. The hoard is the initial line of a set of records in the array is m, the subsequent line in the array is n, and the third line in the array is p. If n [0] = "K", n [1] = "D", and n [2] = "L" then check line is p, else go to step 6, If p [7] = "O" then it is SYN sachet that assaults the system, If computation > 2 then spot the sachet is mistrustful, its further steps are recap stage 2 for the entire file, and Print entire distrustful and other strictures.

Finally, the performance section is about producing the outcome that classifies the data in normal or abnormal classes. The high-quality dubiousness of technologies varies upon better rate rather than the minimum predefined rate. The section reveals the outcome of the analysis if consuming jack screening bout. Doubtful mechanism facts are the total amount of sachets, full semantic sachets, value attained via port count, date, and initial period. The period related to the previous demand reveals a remark in a fixed setup.

5.4   EXPERIMENT AND RESULT

The operation is carried out in the context of efficiently capturing suspicious chunks of packets passing through the network and performing effective preservation and analysis. To implement annexing, and safeguarding of data obtained section, Wireshark is applied from the power shell to get data from a webbing. Evaluation is applied consecutively by customers' outlined requirement on the shell using tools of Mapper for networking, that have some predefined a keyword and "detail" comprises facts of entire sachets. A minor outfit of the dual structure of the machine individually consumes running work. First the IP address 192.168.0.100 is utilized to take off port screening of bout on the target machine associated with the IP address 192.168.0.102 as shown in Figure 5.2.

The above figure divulges the process, depicting the need of capturing data from a network, storing, and analyzing the data for which there is a need for certain tools such as Network Mapper (Nmap) and Wireshark. Consequently, there is an absolute field for examining DT communique and cyber forensic. To instigate the assembling, and salvation section that Wireshark brings into play, it uses of power shell to attain data from the complex lattice. The considered section is applied through the customer's outlined data on the shell of a Nmap tool, where the keyword "IP detail" comprises data of entire sachets. A minor structure of dual machines individually consumes the running system, and a single system having the IP address 192.168.0.100 is utilized to take off port screening of bout on the target machine consisting IP address 192.168.102.

The above table contains numerous commands to perform network scanning and port discovery that is a prerequisite of packet capturing data and analyzing network forensic analysis. Below this, at each, the individual command is consumed to experiment.

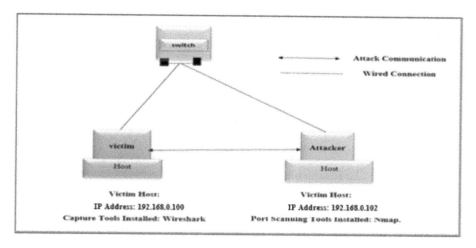

FIGURE 5.2 Attack perform by a single attacker.

TABLE 5.1

Showing Commands for Nmap Tools to Perform Network Scanning and Port Discovery

Action	Port
Perform a Scan	Nmap [target]
Perform numerous Scan	Nmap [target][target]
Port Discover: Don't Ping Command	#Nmap [target]
Port Discover: Ping only Scan	Nmap –sP [target]
ARP Ping	Nmap –PR [target]

5.4.1 NETWORK SCANNING

5.4.1.1 Nmap

The authors (Bagyalakshmi, G., et al. 2018) explained about a tool that is freely available in the market to use is known as Network Mapper (Nmap) and it is used to analyze the packet seized via the lattice of complexity in transmission. Functionally, it discovers a host over a network by constructing a map of the lattice. The tools practice chunks of data toward deciding of many traits related to the data attained from a complex lattice of the network such as accessible route, availability of resources offered, functioning scheme, and control tactics are available with others. The tools for network analysis (Nmap) are obtained in the pattern of GUI named Zen map; otherwise, it can run on the command prompt of Windows, and Linux.

Scan a Single Target: includes performing Nmap with veto command line choice that will make a simple scan on the stated goal and a goal can be stated as an IP address or a hostname (that Nmap will effort to settle).

Usage syntax: Nmap 192.168.0.100
Output

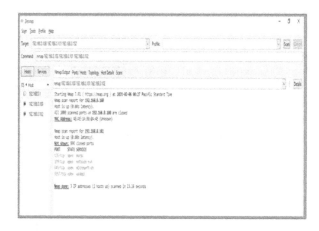

FIGURE 5.3   Single Target Scan.

The above figures unveiling command line choice that will make a simple scan on the stated goal and a goal that can be stated as an IP address or a hostname (that Nmap will make an effort to settle). The command is targeting a single target with the IP address 192.168.0.100.

5.4.1.2   Scan Numerous Targets

Nmap can be consumed to inspect numerous hosts at a constant instant. The usual method to do this is to chain collectively the target IP addresses on the command line (disjointed by a distance).

Usage syntax: Nmap 192.168.0.100, 192.168.0.101, 192.168.0.102
Output

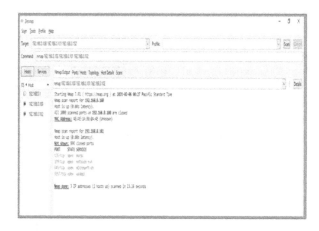

FIGURE 5.4   Numerous Target Scan.

Above figures unveiling command line choice will make a simple scan on the stated targets and there are numerous targets to be scanned. The command is targeting a three target with IP addresses: 192.168.0.100, 192.168.0.101, 192.168.0.102, respectively.

5.4.1.3   Port Discovery

Instead of port scanning a goal, Nmap will effort to direct ICMP echo invitations to find if the host is "available." This can prevent time from wastage while examining numerous hosts as Nmap will not spare time bidding to inquiry hosts that are not connected. Since ICMP invitations are frequently jammed by the control system, Nmap will also effort to link to ports 80 and 443 because these shared web server ports are regularly exposed (if ICMP is not exposed).

5.4.1.4   Port Discovery Command: "Don't Ping"

By default setting, despite the effort to examine a device for exposed ports, it will initially ping the goal to address it is connected or not. This characteristic aids and conserves time during an examination.

Usage syntax: #Nmap 192.168.0.102

Output

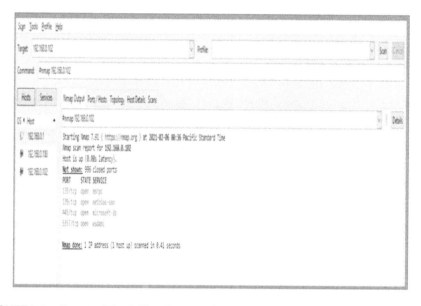

FIGURE 5.5   Output of Don't Ping Command.

Above result revealing port discovery instead of port scanning a target, Nmap will effort to direct ICMP echo invitations to find if the host is "available." This can prevent time from wastage while examining numerous hosts as Nmap will not spare time bidding to inquiry hosts that are not connected. Since ICMP invitations are frequently jammed by the control system, Nmap will also effort to link to ports 80 and 443 because these shared web server ports are regularly exposed (if ICMP is not exposed).

5.4.1.5   Port Discovery Command: "Ping Only Scan"

The -sP selection is consumed to act as a modest ping of the stated host.
   Syntax: Nmap –sP
   Output

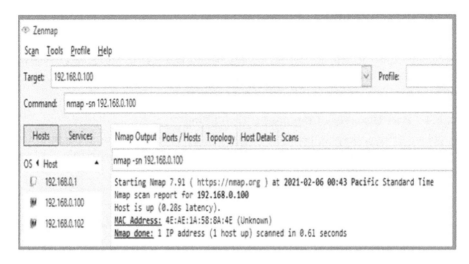

FIGURE 5.6   Ping Only Scan.

Above result gives information about the port discovery by ping command instead of port scanning a target, this helps in preventing the time consumed in scanning. This can prevent time from wastage while examining numerous hosts as Nmap will not spare time bidding to inquiry hosts that are not connected. Since ICMP invitations are frequently jammed by the control system, Nmap will also effort to link to ports 80 and 443 because these shared web server ports are regularly exposed (if ICMP is not exposed).

   ARP Ping: The address resolution command: "ARP Ping". The -PR choice commands Nmap to execute an ARP ping on the particular goal. The ellipsis ARP views for Address Resolution Protocol which is one of the greatest significant etiquettes of the Network layer amongst the layers.

   Usage syntax: Nmap -PR 192.168.0.102
   Output
   Above figure elucidating the outcomes of the address resolution command: "ARP Ping". The -PR choice commands Nmap to execute an ARP ping on the particular targets as 192.168.0.102. There are complete 131,010 TCP/IP ports in a combination of 65,535 with TCP and 65,535 with UDP. Nmap, by pre-set, only examines 900 of the generally used ports. This is made to reduce the wastage of time during examining numerous entities as the popular ports are external and the top 900 are infrequently consumed. Several times, though, we may need to examine externally the pre-set variety of ports to find rare services that have been redirected to a different site.

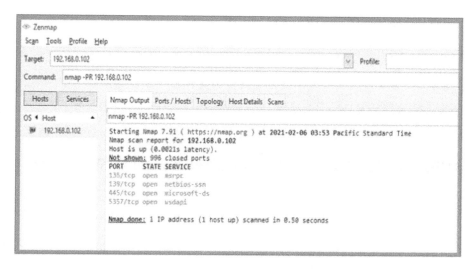

FIGURE 5.7 ARP Ping.

5.4.2 Viewing and Analyzing Packet Contents by Wireshark

5.4.2.1 Packet Capturing by Using Wireshark

Wireshark is a network packet analyzer tool that attempts to capture network packets and efforts to show that sachet information features. This framework is denoted as a network protocol analyzer. The authors (Ndatinya et al. 2015) found that the Wireshark is consumed to scrutinize the aspects of traffic at a diversity of stages varying from correlation equivalent material to the bits that build up a packet. Packet seizure can deliver data about specific sachets such as broadcast time, address of the source, address of a destination, type of protocol, and data of header. These data are valuable for assessing security measures and managing network security machine matters. This exposed-cradle etiquette analyzer is extensively acknowledged as the organization's orthodox, persuasive its impartial part of prizes over the years. At first, named Ethereal, Wireshark has a customized edge that can show information from numerous diverse etiquettes on every main type of network. Data chunks can be observed in inspection online. Wireshark maintains lots of trace file formats maintained comprising CAP and ERF.

Step 1: Choose more than one network by connecting on assumed varieties as WIFI and clunk on the "Blue Shark" Fin key on distant leftward of the toolbar to turn to catch sachets on the network of WIFI.

The above figure shows the packet capturing over a connected Wi-Fi connection using the Wireshark. The above figure exposing the information about numerous packet capturing simultaneously over an IP address 192.168.0.100 by using the Wireshark. The packet captured via "TCP" is showing the source and destination address information. Through source and destination address, analysis of the packet can be done in an easy way.

FIGURE 5.8   Capturing packets over a connected Wi-Fi connection.

5.4.2.2   Using Wireshark Filters

Step 2: Utilize Demonstrate Filter by entering the legal filter term in the demonstration filter unit underneath the toolbar. Every sachet with keyed input filter features will be shown with their particular sachet list, sachets features pieces, and sachets bytes' piece on the monitor. Here in this instance, the Keyed Input Demonstrate filter is "TCP" which will seize sachets having TCP etiquettes.

The above figure shows a packet of **192.168.0.100** IP address is captured by TCP by using the Wireshark. The packet captured via "TCP" is showing the source and destination address information. Through source and destination address, analysis of the packet can be done easily.

The above figure shows the Layered Architecture of a captured packet of **192.168.0.100** IP address in the Packet Details pane by using the Wireshark. The above figure shows a packet of **192.168.0.100** IP address is captured by TCP by using the Wireshark. The packet captured via "TCP" is showing the source and destination address information. Through source and destination address, analysis of the packet can be done easily.

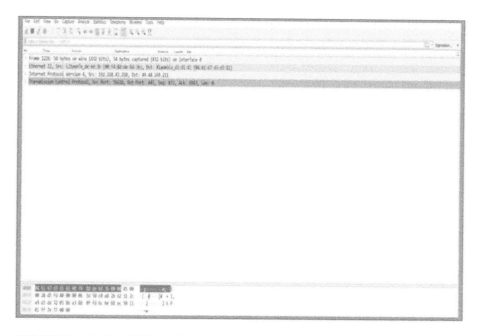

FIGURE 5.9   Packet of **192.168.0.100** IP address is captured of TCP.

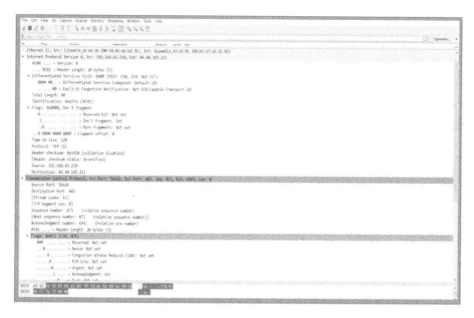

FIGURE 5.10   Architecture of a captured packet.

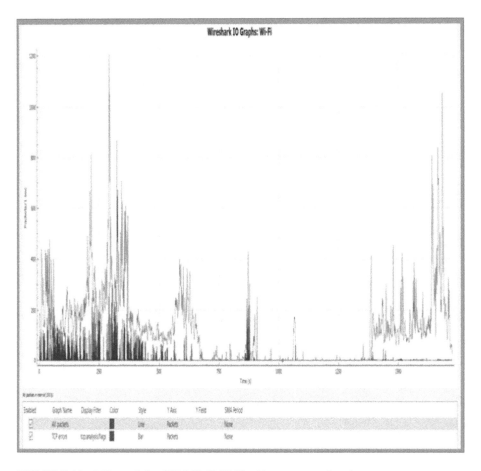

FIGURE 5.11 I/O graph for **192.168.43.77** IP address captured packet.

Step 3: Choose a sachets features pane that reveals the etiquettes and etiquettes fields of the nominated sachets in an inflatable format.

The above figure shows the I/O graph for the **192.168.0.100 IP** Address captured packet by using the Wireshark. The above figure shows a packet of **192.168.0.100** IP Address is captured by TCP by using the Wireshark. The packet captured via "TCP" is showing the source and destination address information. Through source and destination address, analysis of the packet can be done easily.

The above figure shows a packet of **192.168.0.100** IP Address is captured by TCP by using the Wireshark. The packet captured via "TCP" is showing the source and destination address information. Through source and destination address, analysis of the packet can be done easily.

The above figure shows a packet of **192.168.0.100** IP Address is captured by HTTP by using the Wireshark. The packet captured via "HTTP" is showing the source and destination address information. Through source and destination address, analysis of the packet can be done easily.

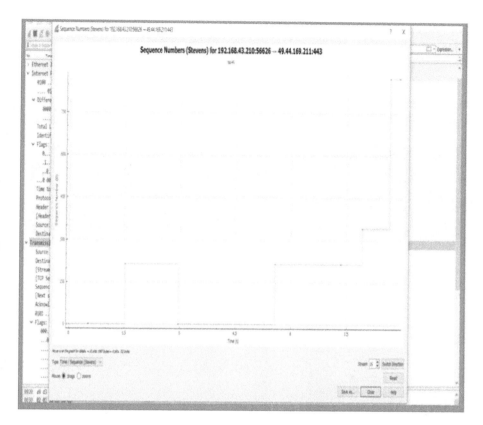

FIGURE 5.12 Sequence number graph.

The above figure shows the I/O graph for **192.168.0.100** IP Address captured packet for HTTP. The above figure shows the Packet of **192.168.0.100** IP Address is captured for HTTP by using the Wireshark. The above figure shows a packet of **192.168.0.100** IP Address is captured by HTTP by using the Wireshark. The packet captured via "HTTP" is showing the source and destination address information. Through source and destination address, analysis of the packet can be done in an easy way

The above figure shows the Non-QoS enabled packets captured to the trail TCP stream by using the Wireshark. The above figure shows a packet of **192.168.0.100** IP Address is captured by TCP by using the Wireshark. The packet captured via "TCP" is showing the source and destination address information. Through source and destination address, analysis of the packet can be done in an easy way

The above figure shows the Non-QoS enabled packets to trail the UDP stream by using the Wireshark. The above figure shows the Non-QoS enabled packets captured to the trail UDP stream by using the Wireshark. The above figure shows a packet of **192.168.0.100** IP Address is captured by UDP by using the Wireshark. The packet captured via "UDP" is showing the source and destination address information. Through source and destination address, analysis of the packet can be done in an easy way

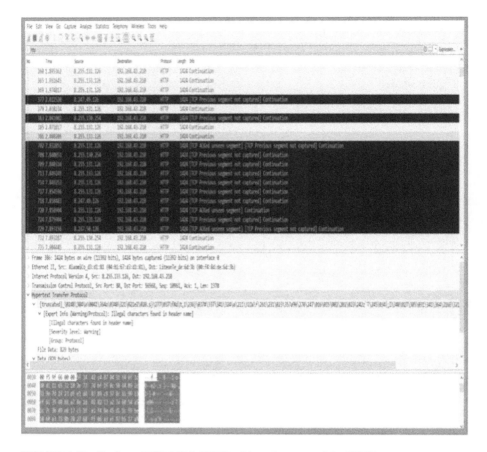

FIGURE 5.13   Packet of **192.168.0.100** IP address is captured for HTTP.

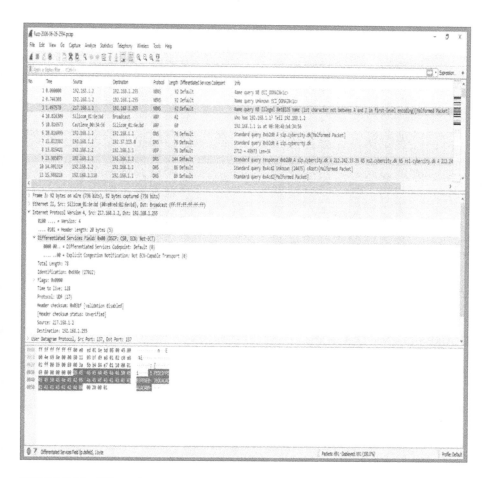

FIGURE 5.14 I/O graph.

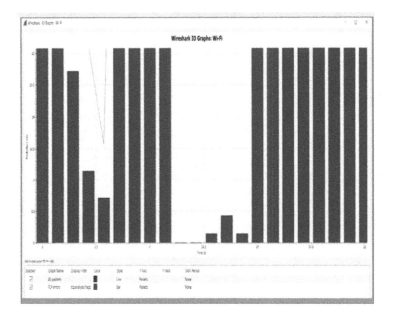

FIGURE 5.15   Non-QoS packets.

FIGURE 5.16   Non-QoS packet follows UDP stream.

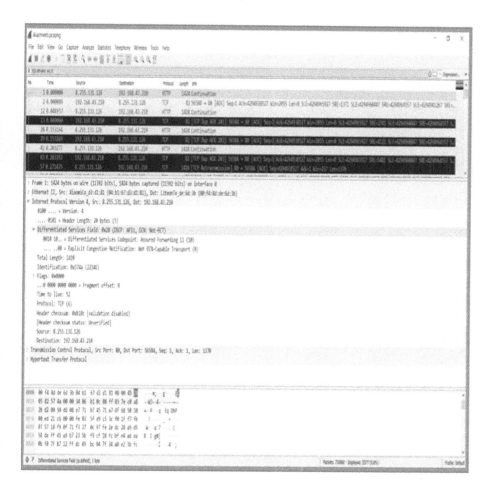

FIGURE 5.17 QoS Enabled packets.

The above figure shows the QoS enabled packets to trail the TCP stream by using the Wireshark. The above figure shows a packet of **192.168.0.100** IP Address is captured by TCP by using the Wireshark. The packet captured via "TCP" is showing the source and destination address information. Through source and destination address, analysis of the packet can be done in an easy way.

The above figure shows the QoS enabled packets to trail the TCP stream by using the Wireshark. The above figure shows a packet of **192.168.0.100** IP Address is captured by TCP by using the Wireshark. The packet captured via "TCP" is showing the source and destination address information. Through source and destination address, analysis of the packet can be done in an easy way

FIGURE 5.18   QoS Enabled packets to follow TCP stream.

5.5   OPEN CHALLENGES IN WIRESHARK AND NMAP

The research gap divulged a feeble portion of the research that was encountered by the authors for developing NFA and showed the limitations of the research.

- The researchers are incapable to use closed access tools and system that is a requirement to the trial to evaluate the result.
- The exclusion of closed access journals destitute the authors of learning from significant and domain-specific chapters of esteemed journals.
- The research made in a specific context by frequent methods closes the doors of bequeathing the fallouts in an improved method.
- The Wireshark is limited in scope because it cannot customize as per the requirement of the situation of network capturing and analysis.

- The Nmap consists of a decent tool to perform the analysis of detected packets and for determining the origin of the source usable only by the authorized personality or government authority.

5.6 CONCLUSION AND FUTURE SCOPE

The notion of Digital Twin (DT) has arisen to allow the welfare of upcoming archetypes such as digitalization in forensic analysis. The persistency of data has crafted an opportunity to create a potential framework similar to DT. The ideas of DT have arisen to style the corporeal entity and associated facts reachable software and customers over digital platforms. This study suggests a framework of forensic subdivision class as forensic analysis via a network to detect bouts, and it is providing well-organized protection. In this chapter, authors are utilizing a self-embedded watermark that delivers reliability, with the existence of waterline, entails a haven code that offers validity to confirm the effectual protection by utilization of many available informant port scanning predefined framework or code included as Nmap, Wireshark, and Advance port scanner. Operation is utilized for evaluating effective salvation and examination. The process needs capturing the data from the network, storing the data, and analyzing it for which there is a need by certain tools such as Nmap and Wireshark and consequently there is an absolute field for examining DT communique and cyber forensic. To instigate the assemble, and salvation section, the Wireshark brings into play the use of a power shell to attain data from the complex lattice. The considered section is applied through consecutive customers' outlined data on the shell of a Network Mapper tool, where the keyword "IP detail" comprises data of entire sachets. A minor structure of dual machines individually consuming the running system consists of the IP address 192.168.0.100 is utilized to takeoff port screening of bout on the target machine consisting of the IP address 192.168.0.102 Future research in the following area:

- The future work includes the use of a Wireshark for investigating the data captured from the network and the use of optimal attribution to train the classifier by machine learning.
- The research will be based on the blockchain supply method that contains a chain of entire information and updates the information in databases regularly as new malware occurs.

REFERENCES

Bagyalakshmi, G., Raj Kumar, G., Arunkumar, N., Easwaran, M., Narasimhan, K., Elamaran, V., & Ramirez-Gonzalez, G. 2018, 'Network vulnerability analysis on brain signal/image databases using Nmap and Wireshark tools', *IEEE Access*, 6, 57144–57151.

Casey, E., & Palmer, G. 2004, 'The investigative process', In Casey, E. (ed.), *Digital Evidence and Computer Crime*, Elsevier Academic Press, 5, 1–14. New york

Chandran, R. 2004, *Network Forensics. Know Your Enemy: Learning about Security Threats*, Second Edition, L. Spitzner (Ed.), Addison Wesley Professional, 281–325.

Kao, D. Y., Wang, Y. S., Tsai, F. C., & Chen, C. H. 2018, 'Forensic analysis of network packets from penetration test toolkits', In *20th International Conference on Advanced Communication Technology*, 363–368, IEEE.

Kaushik, A. K., Pilli, E. S., & Joshi, R. C. 2010, 'Network forensic system for port scanning attacks', In *2nd International Advance Computing Conference (IACC)*, Thapar University, Patiala, India, 310–315, IEEE.

Kumar, V., Singh, A. P., Rai, A., & Wairiya, M. 2011, 'Wairiya: Self alteration detectable image log file for web forensics', In *International Journal of Computer Applications*, 975, 8887

Mandia, K., & Procise, C. 2004, *Incident Response and Computer Forensics*, New York: Osborne McGraw-Hill, 1–11.

Ndatinya, V., Xiao, Z., Manepalli, V. R., Meng, K., & Xiao, Y. 2015, 'Network forensics analysis using Wireshark', *International Journal of Security and Networks*, 10(2), 91–106.

Palmer, G. 2001, 'A road map for digital forensic research', In *Proceedings of 1st Digital Forensic Research Workshop*, Utica, New York, 27–30, LNCS Springer.

Pilli, E. S., Joshi, R. C., & Niyogi, R. 2010, 'Network forensic frameworks: Survey and research challenges', In *Digital Investigation*, 7(1-2), 14–27.

Mandia, K., Prosise, C., & Pepe, M. (2003). Incident response & computer forensics (Vol. 2). New York: McGraw-Hill.

Reith, M., Carr, C., & Gunsch, G. 2011, 'An examination of digital forensic models', In *International Journal of Digital Evidence*, 1, 1–6.

Singh, A., Venter, H. S., & Ikuesan, A. R. 2020, 'Windows registry harnesser for incident response and digital forensic analysis', *Australian Journal of Forensic Sciences*, 52(3), 337–353.

Yasinsac, A., & Manzano, Y. 2010, 'Policies to enhance computer and network forensics', In *IEEE Workshop on Information Assurance and Security*, United States Military Academy, West Point, New York, 289–295.

6 Wind Catchers as Earth Building

Digital Twins vs Green Sustainable Architecture

Hamed Niroumand
a Post-Doc, Assistant Professor, Department of Civil
Engineering, University (IKIU), Qazvin, Iran
and
b Post-Doc, Powell Center for Construction and
Environment, University of Florida, USA

Charles J. Kibert
University of Florida

Somayeh Asadi
Pennsylvania State University

Fatemeh Mahdavi
Tehran University of Art

Hadi Arabi
Pars University of Architecture and Art

CONTENTS

DOI: 10.1201/9781003132868-6

6.1   INTRODUCTION

Sustainable development is a concept that addresses the requirements of the present, as well as the next generation and considers future exigencies that need to be met. In fact, this implies the priorities are assigned as per the world demand, and taking into consideration the constraints, that technology and social organization impose on the present and future needs [1].

Sustainable buildings are structures with maximum use of materials along with minimum resource consumption while ensuring the health and well-being of residents and the environment of today and future generations. LEED's* instruction addresses seven topics in the design and construction of new eco-friendly buildings that include (1) sustainable site, (2) water efficiency, (3) energy and atmosphere, (4) materials and resources, (5) internal environment quality, (6) innovation in design, and (7) regional priority [2].

Sustainable architecture is also known as green architecture or green building, and its most important approach is maintaining and preserving the environment and designing buildings in the local ecosystem and the global environment [3]. In the past, architects have developed a series of creative strategies in the buildings; for example, they use only natural ingredients, that don't pollute, are cost-effective, and are maintenance-free materials [4]. As in the Green Energy Division in the LEED instruction, it is mentioned that the goal of sustainable design is the encouragement of the use and development of Grid Source. It is the renewable energy technology with zero contamination [2].

Wind catchers are considered for ventilation of residential houses. The installation of a suitable ventilation device, especially in hot climates, has been an important requirement for over the last years. Today, the wind catcher is still a comprehensive and international solution for air conditioning in the warm climate regions [5]. Wind catchers have always been one of the main components of buildings in the central regions of Iran and neighboring countries. The wind catcher is a protrusion that draws the wind from the outside air and enters into the building [3]. The wind catcher can also be referred to as a passive cooling system that reduces energy consumption, CO_2 emissions, and contamination. This system can eliminate undesirable heat pollution from the building and provide thermal comfort to the residents [6]. In the LEED Guide, the indoor environmental quality part (IEQ) in the Energy and Atmosphere Division, it points out that the goal is to increase air quality of the building in a way to provide comfort and well-being of the residents [2].

Natural ventilation drew the attention of builders to reduce energy consumption and launched a new study series called low energy architecture [7,8]. It can be said that LEED's book in this regard has stated that the goal of having buildings with minimum energy efficiency is to reduce the environmental and economic impacts associated with excessive energy consumption [2].

Indoor air quality (IAQ) is one of the top five environmental threats to global health and well-being, and the research community has discovered the scope of artificial intelligence (AI) in recent years to address this problem. IAQ prediction systems help intelligent environments, and advanced sensing technologies can create healthy living conditions for building occupants. [9] Housing has not changed much in terms of typology and performance since the early 20th century, yet meticulous work in computing, the Internet of Things (IoT), and green architecture seek to bring technological innovation to future homes through the idea of smart homes. Of course, smart does not always mean sustainable. One of the tasks that can be done in smart homes by artificial intelligence is to install natural ventilation by smart wind catchers that are able to position themselves in the right wind direction and have optimal performance. In fact, traditional structures with centuries of life can be used intelligently. Smart wind catchers are passive designs because they use environmental resources (wind) to ventilate the interior and so enable passive cooling. The more the building relies on passive systems, the more sustainable it is. Technology makes traditional structures flourish and eliminates their shortcomings. Using technology and Internet of Things (IoT), you can control the amount of wind entering the wind catcher, but the problem with artificial intelligence (IA) and intelligent systems is that it needs to be maintained, updated, and fixed, and after a while, there may be problems and it loses its effectiveness [10]. However, the use of technology to increase the efficiency of natural ventilation can be significant and useful in many ways, and the quality of naturally ventilated air is 40% higher than the quality of air ventilated by ventilation devices [11].

In this paper, a comprehensive overview of historical wind catchers around the world and their origins are investigated. Modern wind catchers in different parts of the world have also been studied, and an attempt has been made to find a comprehensive article, covering all aspects of different types of wind catchers in different locations, as well as presenting specialized analysis on the machining and improvement of the performance of wind catchers, as studied in various researches.

6.2 PASSIVE VENTILATION AND NATURAL VENTILATION

Passive here refers to devices that do not have a moving part, although some of the latest improvements and innovations have implemented the capability of moving and controllable progress has been made. A passive way to provide the desired air and comfort without the need for mechanical systems is through natural ventilation which uses sustainable and energy efficient methods to ventilate indoor air [12]. The ventilation of the building, both mechanically and naturally, plays different roles in the indoor air quality and promotion of the level of thermal comfort in the summer and the reduction of energy consumption is based on climatic conditions, but the natural ventilation advantage is the exploitation of a free and abundant source that is easy to use. So, natural ventilation is a very good solution to the current problem of designing low-energy buildings with the lowest greenhouse gas emissions [13].

Passive cooling techniques include various methods, such as ventilation cooling, solar chimney, evaporative cooling, ground cooling, night ventilation, indirect evaporative cooling, and indirect radioactive cooling [14,15]. Among these techniques, ventilation cooling has a subset called passive ventilation, which includes various methods, such as pop-up windows, atrium and courtyards, wing walls, chimney cabins, and wind catchers. Our focus is on passive ventilation of wind catchers. The wind catchers have different divisions. In this study, we divide the wind catchers into two types of native and modern [14].

6.3 THE STRUCTURE AND MECHANISM OF TRADITIONAL WIND CATCHERS

Pirhayati et al. [16] introduced two types of wind catchers in their study. First, wind catchers that cool the interior space during heat transfer. This type of wind catcher is used in warm and humid areas such as Bushehr, Lar, etc. The second type of wind catchers are those that use evaporation and heat transfer to cool the inside area.

Mahmoudi Zarandi [17] has been studying on Yazd and Lengeh port wind catchers and stated that Yazd wind catchers should pass through a moist surface and should not reach the main room directly, but the wind catchers in Lengeh port directly enter the air from wind catcher to the room without the need for any moist surface. She also states that the height of the wind catcher increases its function. The height of the wind catchers are almost 5 m in the Yazd city and 3.5 m in the Lengeh port city. The area of most wind catchers in Yazd is on average between 3 and 5 m², and the average area of the Lengeh port wind catchers is 9 m².

In general, the behavior of airflow in a building depends on natural environment of that area and can be predicted by the following equation [18]:

$$\frac{1}{2}pv^2 + P + pgh = \text{constant}$$

p: Density (kg/m³)
g: Gravitational acceleration (m/s²)
v: Speed (m/s)
h: Height (m)
P: Pressure (Pa) [18]

In Figure 6.1, the concept of airflow has been shown schematically in traditional Iranian houses.

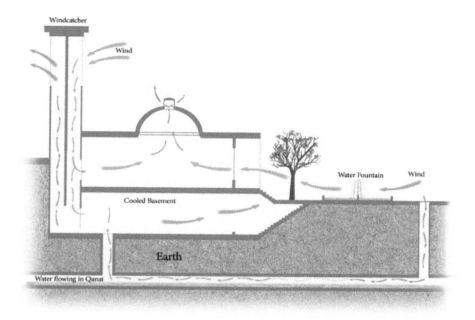

FIGURE 6.1 Air flow direction in traditional Iranian houses.

Yaghoubi et al. [19,20] discussed the details of the wind catcher mechanism. They concluded that high-rise wind catchers, such as the Lary and Dowlatabad House in Yazd are heated due to exposure to sunlight and do not function properly. An innovative solution at that time was the passage of air from an underground channel called Naghab, which increases humidity, to make evaporative cooling happen. Usually, the channels that the air passed through, traversed the gardens at depths of 5–7 m, and the water used to irrigate the plants would also moisten the channel. In Figure 6.2a and b, the structural details of the channel are shown; thus, the air enters through the openings and after passing through the channel, it enters into the underground part of the Ameri house. The channel is tilted and steep because the cooler air is heavier than the hot air and it causes cold weather to go down and enter the house. This type of channel, even if the wind does not blow, replaces the hot air of the house with cold air, but if the wind blows, the airflow rate is higher and the channel's function is even greater.

Pirnia [21] writes in the book *Islamic Architecture of Iran* that in some cities such as Kashan and Tehran houses had two wind catchers. The second wind catcher was built on the side of the alcoves. In this type of wind catchers, the first tower is the wind turbine and the second tower works as a ventilator. He also writes that wind catcher mechanism, especially in the warm and central cities and around the desert, has been so fundamental and conservative that it can be argued that science and technology today, with all its progress and development, has not been able to make a better alternative to it.

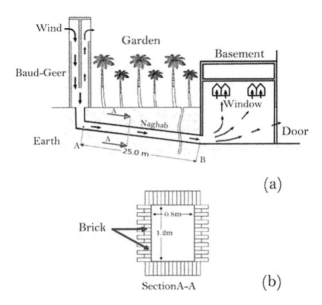

FIGURE 6.2 (a) Schematic image of the airway through the wind catcher and channel at the Ameri house. (b) The section of the channel [19,20].

6.3.1 CHIMNEY EFFECTS

The chimney effect is simply understandable with attention to the interior space of a building as a column of air that connects two valves at a z-spacing. We assume that the interior spaces have a mean temperature of T_i and a medium concentration of p_i. In the surroundings of the building, there is a volume of air in which the temperature at the bottom of the building is lower than the average outside temperature of T_o, and therefore its concentration is higher. The internal air column between the valves inserts the pressure p_i (z) on the horizontal surface that passes through the center of the lower valve, while the external air with equivalent height imports the pressure p_o (z) to this level. So, the outside air enters due to heavy loads and replaces the light air inside, which goes out from the upper hatch. The difference in pressure (the Delta P chimney) which drives the air is called the "the pressure of chimney" or "flow of air", and it is the difference between the forces that gets exerted by the internal and external air columns with the height of (z) [22].

Delta P chimney $= (p_o - p_i)z$ Delta P chimney $=$ The pressure difference that drives the air (psf)

$Z =$ Vertical distance of air hatches (ft)

$p_o =$ Concentration of external air (PCF)

$p_i =$ Concentration of internal air (PCF) [22]

FIGURE 6.3 Several examples of traditional Yazd wind catchers.

6.3.2 WIND CATCHER EFFICIENCY

Montazeri and Azizian [3] conducted a study in Yazd, a place known as the city of wind catchers, to analyze the number of wind catcher openings and their functions. In Figure 6.3, some examples of traditional wind catchers in Yazd are shown. These studies have shown that the efficiency of wind catcher depends on height, cross-section and, the location of the wind catcher, as well as the number of openings it has. Also, the angle of the inflow of the wind and the presence of surrounding buildings, along with general flow of the winds affect the coefficient of pressure, the amount and direction of the air conditioning flow.

Mahmoudi Zarandi [23] has obtained different results by examining three different models of wind catchers in the studied house. The type of architecture and direction of the wind catcher is important in the behavior and the heat performance of the wind catcher, although the plans of the houses, installation of the wind catcher on the house, and the wind catcher's height and the fittings are uniform. It was concluded that the best performance among the various wind catchers belongs to the wind catcher that are in + shape. Wind catcher dimensions are calculated with respect to external air temperature. If the temperature of the inflow air is low, the horizontal cross-section should be large; but if the inflow air temperature is high, then the horizontal cross-section must be small. Also, wind catchers should be placed in internal walls to avoid exposure to sunshine and moisture absorption.

6.4 THE HISTORY OF WIND CATCHER

Wind catchers have always been defined as a traditional ventilation structure and have been used with different names and forms in the Middle East, from Pakistan to North Africa [14]. Wind catchers have been used in countries with hot weather for centuries. As their name suggests, they are ventilation devices used for natural cooling. Wind towers in the central cities of Iran are known as "Badgir", the literary meaning of which is wind catcher. Wind catchers are not unique to ordinary houses, but also seen above the water reservoirs and mosques. The earliest historical evidence of the wind catcher reaches the fourth millennium BC. An example of a very old wind catcher in Iran was found by a Japanese explorer in a house on the Chakhmakh Hill, about 8 km north of Shahrood, on the slopes of the Alborz

Mountain range in northeastern Iran [24]. The wind catcher has been used for centuries to make ventilation possible in the building. It is unclear exactly who the inventor was, but different types of wind catchers in different parts of the world, from ancient Egypt in the houses of Tal al-Amarnah and in the Pharaonic house of Neb-Amun (from the 19th century 1300 BC) to The Middle East and Europe have been seen [6].

Pirhayati et al. [16], in their research on wind catcher history, have come to the conclusion that the invention of wind catcher dates back to the time of Christ, but it is hard to say which country first invented the wind catcher. The Egyptians, like their famous architect Hassan Fathy [25], according to the paintings from the tomb of Neb-Amun in Tal al-Amarnah in 1300 BC, know the source of the wind catcher in Egypt, but according to Masuda explorations dating back to 4000 BC, the assumptions of researchers such as Hassan Fathy and Abdul Moonim al-Shurbeki are invalidated. The Arabs have used the Persian word "Badhenj" in their poems, and today the word "badgir" is used by the Arabs, which indicates the history and background of the wind catcher in Iran. In addition, the first wind catcher was constructed in the United Arab Emirate (UAE) in a regional bay near Dubai, the first inhabitants of which were Iranians living there. Therefore, given the results obtained, it can be said that Iranians were the inventor of this masterpiece, and, like several other architectural elements, this element of architecture has been sent from Iran to other countries and has been the source of their inspiration [16].

Figure 6.4 shows an example of indigenous wind catchers and lightning on the roof view and the interior view of the dome that belongs to the historic Boroujerdi House in Kashan, Iran.

FIGURE 6.4 An example of indigenous wind catcher and lightning in the dome and roof view of Boroujerdi houses, Kashan, Iran.

6.4.1 IRANIAN WIND CATCHERS

The creation of a wind catcher is the answer to the understanding of Iranian architects of the phenomena of nature and wind, and it can be said that the climate has the greatest impact on the architecture and traditional elements of Iranian architecture [16]. The wind catcher has been used since ancient times in Iran, and some of its ancient and varied names are Watghar, Bādhenj, Batkhan, Khishud, and Khishkhan. It is not a new phenomenon, and now all the air-conditioning equipment and its types are based on wind catchers and Khishkhans. Wind catchers in Iran are of different types and have been constructed in different parts of Iran due to difference in climate and wind direction. The most beautiful and prolific of them are around the dry and burnished plains, especially in the cities of Kashan, Yazd, Bam, Jahrom, Tabas, Persian Gulf and Arvand River [21]. Wind catcher is one of the wonders of Iranian architecture that emphasizes the intelligence of the ancients on the coordination of climate, and it can be one of the examples of the clean energy index. Iran has the largest number of wind catchers. These wind catchers are made in two climatic zones: warm and sultry southern regions (such as Lengeh port) and warm and dry areas (such as Yazd) [3].

In addition to the desert houses, wind catchers play a major role in water reservoirs in Iran. Water reservoirs were places where cool water was collected during the winter days to be used during the hot summer months. The role of the wind catcher is to keep the water cool and ventilate the warm air over the tank. During the day, the air underneath the dome of the water reservoir was warmed up, and this hot air was ventilated by the wind catchers and the aperture above the dome. The wind catcher also removes the collected water vapor, so that the reservoir's water can still keep its temperature low. In Figure 6.5, the flow of air is evident in the water tank [26].

Khishkhan is also a means used in ancient times in Iran that spread to neighboring countries and even to remote climates. Khishkhan was a hut which was covered with mat, pottery or thistle and they threw water on it so that the cool air could come

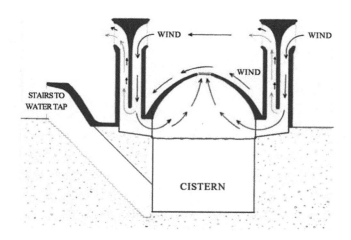

FIGURE 6.5 Ventilation and direction of air flow in the water reservoir [26].

inside, even today's cooler is nothing but a Khishkhan that artificially and with the help of electricity blows the wind [21].

Successful examples of Iranian wind catchers have been used in traditional Kashan houses. Saljoughinejad and Sharifabad [27] studied three traditional houses, Tabatabai, Germany and Boroujerdi, and found that climate strategies have been used in these houses. These multiple strategies work together, each one of them is placed in one location and upgraded each other's performance. These strategies make people feel comfortable in difficult weather conditions. For example, the wind catcher strategy that is considered for a building will improve its application by using other strategies such as domes, lighters, water, and vegetation.

6.4.2 TRADITIONAL WIND CATCHER SAMPLES ALL OVER THE WORLD

6.4.2.1 Badkhan in Pakistan

A simple wind catcher is made from clay, bush, or mat [28]. Heydar Abad wind catchers in Pakistan are at least 511 years old. Pattinger, a British tourist in 1815, said about Heydar Abad: "All houses, state palaces, and small cottages have wind catchers." The wind catcher is stable in its position to reach the prevailing evening winds towards each of the multi-story building rooms. Although this cooling equipment seems to be very different from cooling equipment in other areas, this device is considered as alternative solutions or is an alternative to meet the environmental needs of the area. The design of wind catcher in Pakistan is a square that is surrounded by two vertical sheets and its dimensions are like a square wind catcher in Afghanistan.

The cover of the roof is a ramp with a 45° angle. Wind receivers are covered with wood, plaster, and metal sheets that are placed on this cover in new models. The average wind catcher size is about $1\,m^2$, and its height is more than $5\,m$, and an example of these wind catchers is shown in Figure 6.6. [29].

FIGURE 6.6 Pictures of the traditional Badkhans in Heydar Abad-Pakistan [29].

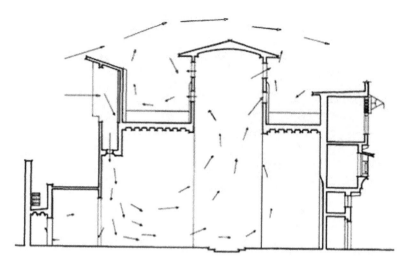

FIGURE 6.7   Section through the Qa'a of Muhib AlDin for wind stream [25].

6.4.2.2   Malkaf in Egypt

Another technique used for natural ventilation is a device called Malkaf, modeled in the ancient Egyptian architecture – the wind catcher which drew the wind to the building. This device is divided into four quarters that allows fresh air to enter, as well as allows the old and used air to leave from the remaining quarters [30]. See Figure 6.7 for the flow of air in a house by a Malkaf.

6.5   WIND CATCHER TYPES IN TERMS OF THE NUMBER OF OPENINGS

Wind catchers are divided into different categories in terms of the number of openings: one-sided, two-sided, multi-directional, and wall wind towers.

A. **One-sided**: One-sided wind catchers are usually made in areas where the wind is predominant [6]. Wind catchers in the coastal areas are only built in the form of a ventilator in the back of the sea and are in contrast to the direction of the wind of the sea. Iranian rectangular wind catchers are made in areas that usually have wind coming from one direction in the summer, and it is usually from the northeast to the southwest; therefore, the big surface of the wind catcher is built right in its face. In the villages next to and within the desert to avoid tornado and heavy storms, wind catchers are built only on the north-east, and other fronts are closed. The roofs of these wind catchers have been constructed in the form of trusses due to the stability factor against the whirlwind [21], which is evident in Figure 6.8, with its plan and section.

B. **Two-sided**: In this type, each opening has a specific application; an opening is located in the direction of the prevailing winds to absorb the desired

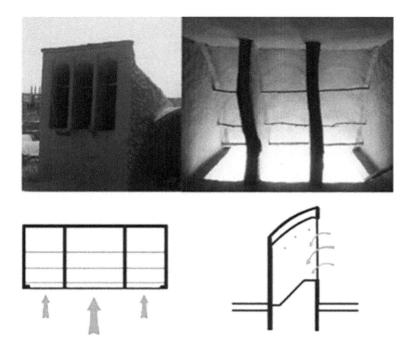

FIGURE 6.8   An example of a one-sided wind catcher and its details.

air; the other opening is in the opposite direction, to direct the warm air absorbed from the room outward [31]. A short rotation is a phenomenon that occurs if the angle of inflow air is greater than zero degrees; a short rotation causes the air to enter from the entrance side and leave from the other, without the opportunity of streaming in the building [6]. Its schematic plane and section are shown in Figure 6.9.

C. **Multi-directional**: This type of wind catcher generally opens in four directions to absorb air as shown in Figure 6.10. But there are a few octagonal and hexagonal wind catchers in the Persian Gulf countries and Iran. There are different designs for this type of wind catcher, but the square shape is more common [31]. Four- and eight-sided wind catchers are suitable for areas that are diverse in wind direction, especially in hot season. Sometimes, the eligible wind flows from the north to the south and from the east to the west [21]. For example, the Dowlatabad eight-sided wind catcher of Yazd is shown in Figure 6.11, which is the tallest wind catcher in the world with 34 m height [32].

D. **Wall wind towers**: These towers are based on the theory of the effect of wind pressure on the wall surfaces in large rooms. Their outer shape is hollow and horizontal pieces are located in the top center of the outer wall. The air flow with high pressure collides with the dominant wind at the outer wall surface, and the air enters the niche and, by pushing it into the openings, creates an inflow of air inside [31].

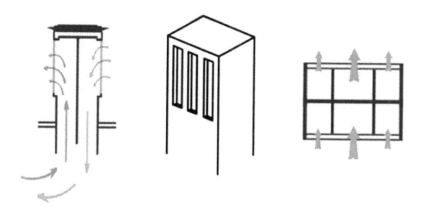

FIGURE 6.9   Plan, three-dimensional schematic and section of the two-sided wind catcher.

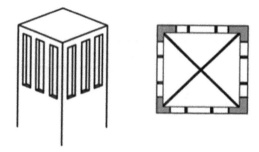

FIGURE 6.10   Plan and three-dimensional schematic of the four-sided wind catcher.

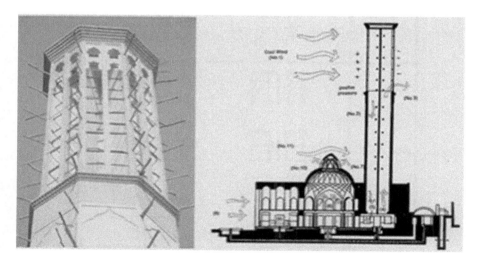

FIGURE 6.11   Eight-sided wind catcher of Dowlatabad in Yazd city [32].

6.6　BLADE LAYOUT AND THE PLAN FORM OF WIND CATCHERS

Wind catchers have different categorizations and classifications. Here, we classify the blade layout and the wind catcher plans:

The wind catchers have triangular, square, circular, and rectangular planes. The triangular and circular planes are very scarce, and the triangular form of wind catcher is not found anywhere in the world except the Middle East [16]. In Figure 6.12, categories of wind catchers in the form of plan are shown.

Blade layouts of wind catchers are of five main types, as follows:

×-**Shaped blades**: The length of this wind catcher is usually 1.5 times its width. This technique has been seen less in the traditional houses.

+-**Shaped Blades**: In this type, the blades are perpendicular to each other and it is the most common type of wind catcher in traditional houses.

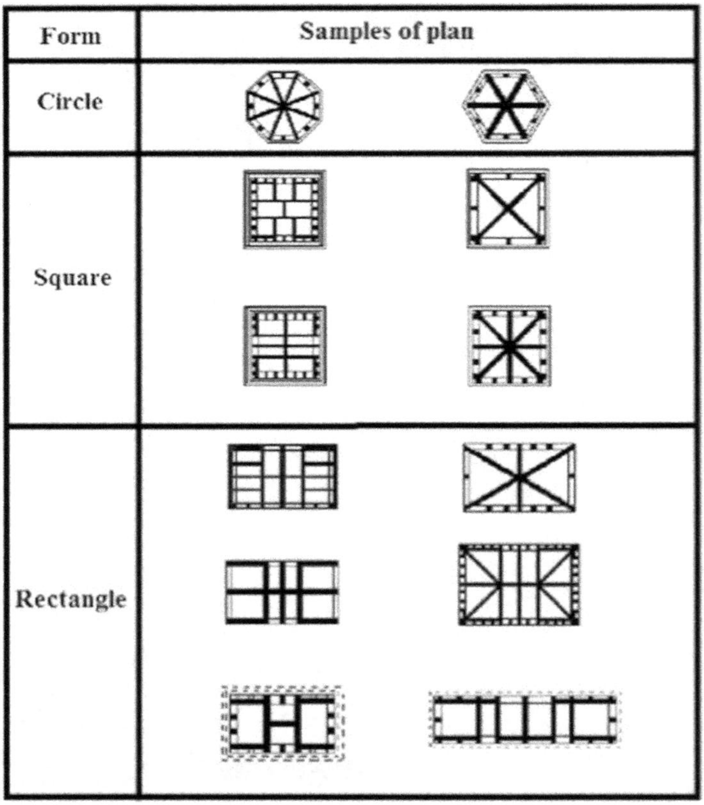

FIGURE 6.12　Types of wind catcher, based on the plan form [16,33].

H-shaped blades: In this method, the main blade separates the wind catcher duct. It is located in the center of the canal and does not extend to the transverse walls of the wind catcher. The symmetric plan is square and cannot be rectangular. This method is rarely seen in traditional houses.

K-shaped blades: This is a combination of ×-shaped and +-shaped blades, and it is rarely seen in traditional houses.

I-shaped blades: The main blades are hidden in the transverse front of wind catcher. There is a closed opening on the opposite side of the open gap to allow the wind to escape. This blade layout method has the widest rectangle [34].

In Figure 6.13, the blade layout types of wind catchers are shown.

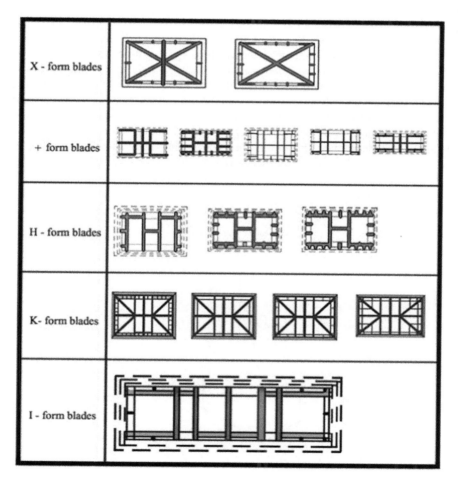

FIGURE 6.13   Plans for different shapes of blade layout inside the wind catchers [34].

6.7 MATERIALS, COLOR, AND TEXTURE
OF THE WIND CATCHERS

Wind catcher materials are selected according to the climatic conditions of the region. As a passive cooling system, the material selection is a major factor in ensuring proper operation of the wind catcher. In hot and dry areas, most wind catchers are made of mud, straw, and bricks. The mud that covers the exterior surface of the wind catcher is bright to reflect the rays. In warm and sultry areas, the wind catchers are covered with plaster and sarouj because the moisture penetration to this structure leads to gradual degradation and this material is resistant to moisture. White is also selected for wind catchers in these areas because it does not absorb solar rays [35,36].

In a study conducted by A'zami [32], it was concluded that the main material of the wind catchers in Yazd, which is a warm and dry city, is mud bricks or brick with clay and straw due to the longer time of heat transfer. The main material of the wind catchers in Lengeh port, which has a warm and humid climate, are plaster and lime plaster to prevent the penetration of moisture.

Also, in some of the wind catchers, there are horizontal wooden beams that are used for strength and resistance of wind catchers against the wind, and preventing the separation of blades and the body of wind catcher and the extra scaffolds do not cut off and they are used for restoration of wind catchers [37]. In each geographic area, the indigenous materials of the same area are used, which is one of the main characteristics of sustainability of wind catchers. It can be cited to the materials and resources section of LEED that the purpose of indigenous and regional materials section is to increase the demand for construction materials and products extracted and produced in the local region, which have the result of supporting the use of indigenous resources and reducing the environmental impact of transportation [2]. The shape of the wind catcher blades and its structure are evident in Figure 6.14.

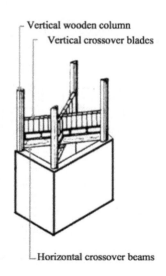

FIGURE 6.14   Structural details of the blade layout [38].

6.8   DISADVANTAGES OF TRADITIONAL WIND CATCHERS

Wind catchers that were used in traditional houses were not perfect. In some traditional houses, these wind catchers have caused unpleasant conditions. Natural ventilation can have its own disadvantages and problems, such as the entry and exit of sound and insects, the limited cooling potential of wind catchers in the summer, humidity, night cooling, and safety [39]. The wind catchers have problems such as the entry of dust, birds and the impossibility of controlling the humidity, airflow, and refrigeration [37]. Also, the wind speed sometimes comes down very low and wind catcher does not work at a low wind speed, taller wind catchers work better, but because of the structure and architectural issues in the traditional buildings, they cannot be very tall [4,35].

6.9   THE CHARACTERISTICS OF USING MUD AND STRAW AND THE POSITION OF MUD ARCHITECTURE IN SUSTAINABLE DEVELOPMENT AND GREEN BUILDING

In the area of Mesopotamia in very distant times, clay was one of the main materials used in buildings [40]. Also, Kahgel (Persian idiomatic term consisting of two words Kah means straw and Gel means mud) is one of the oldest traditional Iranian mortars, which due to its capabilities in the past, was used as an appropriate cover for the protection of the soil architectural structures, but one of the problems to be considered is the need for Kahgel to be repaired and rebuilt against natural erosion, caused by atmospheric agents [41].

Clay is a mixture of soil, straw, and water, and is a solid and sturdy compound. The mixture is poured into a mold and then exposed to the sun for 2 weeks to dry. Then, they are ready for use in the building. The mortar used is also prepared in the same way. This mortar has a very high adhesion property. The clay is also very resistant at first, but their resistance to crack is reduced over time due to Kahgel decay.

Clay provides better insulation, compared to cooked bricks and concrete bricks. Its environmental compatibility, ease of construction, versatility, and its easy cutting, are some of the important features of this building material. It is also easy to use and does not require high skills. In addition, the clay brick is very durable [40].

Sustainable materials have low energy storage, and their production and processing are environment friendly; they have environmental reversibility possibility with the least negative impacts on the environment. These materials should not be radiant or toxic and should guarantee health of the residents as well as the installer's workers. They should have a high lifespan and low maintenance costs. Extraction, processing, production, refining, and transportation of building materials are all partly responsible for ecological damages [42].

Clay and Kahgel can be a good option for replacing with non-renewable and unstable materials, such as metal and concrete, etc., for corrections of problems, such as longevity and erosion, as well as structural strength. As regards to improving the performance of Kahgel, Masoud Bater et al. [41] conducted researches and made it clear that by adding micronized silicate materials such as microsilica, feldspar, zeolite, bentonite, and kaolin, mechanical and physical properties of the Kahgel can be greatly improved. Measuring the permeability properties and the rate of Kahgel resistance to water erosion using rain simulators, they concluded that the use of 3% of kaolin (150 μm) causes permeability decreases to 65% and adding 3% of zeolite (45 μm) causes 85% of decrease in the permeability and the insulating properties of the Kahgel increased. Also, the use of 3%, by weight, of microsilica, feldspar, zeolite, and kaolin micronized got reduced from 10.5% to 37.7% of the waste solids under artificial rain test. These studies have shown that reducing the particle size of additives increases the positive effect on the improvement of physical and mechanical properties in the Kahgel mortar. Also, the optimal amount of these materials is 3% by weight, and increasing the weight will not significantly improve the condition of the mortar.

6.10 MODERN WIND CATCHERS: DIGITAL TWIN ARCHITECTURE

6.10.1 IMPROVE THE PERFORMANCE OF THE WIND CATCHERS

The evaluation of the performance of modern wind catchers is based on empirical researches, using a wind tunnel test. This is essential for evaluating the performance of wind catchers for natural ventilation in buildings, especially in the large-scale model. The experimental results have a great deal of credibility in the research community, as well as in the industry [43].

Due to high energy consumption in residential areas around the world, the goal of sustainable design is to minimize energy consumption and cost. Chaudhry [44] suggested using HVAC technologies in residential design. The use of an HVAC system is to achieve the same level of ventilation as electricity and mechanical appliances, but with the exception that natural and passive ventilation systems are used here. One of them is the natural ventilation of wind turbines, which are currently running in large numbers. The integration of natural ventilation by wind catchers as an alternative to mechanical HVAC systems or as supporting them has the potential to increase indoor air quality and reduce carbon emissions. The results show that the cooling potential in the Middle East is up to 12°C in the warm climatic conditions. Scientific advances in research on the use of passive HVAC mechanisms for effective transfer of heat gradually increases with the use of CFD codes and sophisticated experimental techniques. In Figure 6.15, a comparison of the cost and power of HVAC systems on the input scale is shown.

Sayed [15,45] studied passive air suction system used in a single bedroom apartment in Assiut, Egypt. The system is combined with a horizontal wind catcher and steep solar chimney to extract air and produce natural airflow. With this system, a palpable change at internal temperature is felt that ranged from 6°C to 7°C, while relative humidity remained at 57%. The maximum amount of air flow in the chimney reached 3.5 m/s. The low cost and high performance of this technology are due to the advantages of using it in high scalability.

Gharakhani et al. [46] studied the performance of wind catchers in tropical regions with wind tunnel tests and CFD simulations. The results showed that increasing the height optimizes the performance of a wind catcher by taking into account other variables, so the studies showed that the proposed system has the ability to ventilate even under low wind speeds. According to the results of wind tunnel tests and CFD

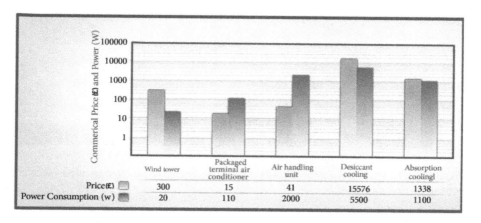

FIGURE 6.15   Comparison of cost and power of (HVAC) systems on the input scale [44].

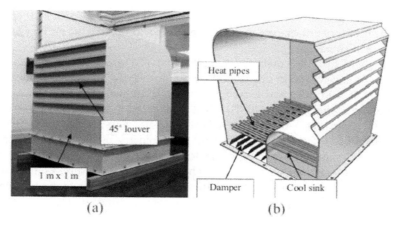

FIGURE 6.16   (a) The wind catcher prototype. (b) A 3D sample which represents the inside of the system [48].

simulations, the wind catcher can be used in a warm and sultry climate in order to provide thermal comfort in green buildings by increasing natural ventilation.

Badran [47] has studied the design and operation of cooling towers, based on climate conditions and comfort criteria in Jordan's Oman, and has concluded that in this climate condition the tower's height, which is necessary to provide proper conditions for the flow of cool air in space, needs to be less than 9 m, and this contrasts with a native design that reaches a height of 15 m. It was also concluded that much fewer heights were needed to create a proper cooling effect. A tower with a height of 4 m and a 0.57 × 0.57m cross section can produce 1 ton of refrigeration.

Calautit et al. [48] conducted experiments in the city of al-Khaimah in the UAE on modern wind catchers. They performed their experiments in a 3 × 3 × 3-m³ test room, depending on the monthly wind speed, wind direction, air temperature and relative humidity by a modern wind catcher, as detailed in Figure 6.16. This modern

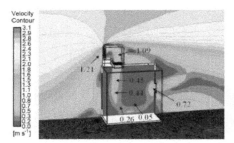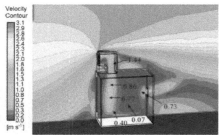

FIGURE 6.17 Distribution of predicted velocity magnitude (m/s) for a wind catcher with (a) without (b) heat pipes [48].

wind catcher contains a number of heat pipes when the hot air passes, the pipes absorb the heat, and then the temperature of the pipes is kept low by a cooling sink. It can be seen in Figure 6.17 that by placing the heat pipes due to variations in the density and air pressure, the wind speed increased in the wind catcher. This research shows the positive performance of the heat pipes in accelerating the air flow and also reducing the temperature.

Dehghan et al. [18] conducted experimental and analytical experiments on two-sided wind catchers. In this trial, three single-sided wind catcher models, with a scale of 1:40 and steep and curved roofs were examined. The experiments were carried out at a wind speed of 10–20 m/s at an angle of 0°–60° with a 15° increase. The measured coefficient of pressure indicates that the pressure induced by the wind catcher openings is heavily influenced by the geometry of the wind catcher roof and the direction of the wind. As the angle of the air increases, the pressure coefficient around all internal surfaces approaches the ambient pressure. Comparison of pressure coefficients also shows that at the high wind angle, the highest difference between internal and external pressure coefficients is observed. The wind absorption coefficient increases with increase in external air velocity and decrease in wind absorption capacity occurs with increasing wind angle. Comparison of semi-analytic results with experimental results shows that theoretical modeling can predict performance of the ventilation of wind catcher for the entire range of wind flow and its direction. But for wind catchers that are affected by the output components, semi-analytic methods are needed to observe the ratio of airflow and ventilation usage and may have an error rate of 16%–18%. In Figure 6.18, the diagrams of this experiment are shown.

Increasing the height of the wind catcher can help to improve efficiency of wind towers. In this case, the wind catcher gets more powerful winds and can lower its height when the wind speed is suitable for lower elevations. Another way is to use cellulose cell at the inlet of a wind catcher with a water tank that increases the moisture content of the wind and decreases the temperature due to moisture. One of the best ways to upgrade the wind catchers is to use the sun's energy by sticking solar panels to the top of the wind catchers. These panels absorb and store the sun's energy. The fan in the wind catcher powered by solar energy starts to move on sunny days when the wind speed is low and can be controlled by moisture-absorbing

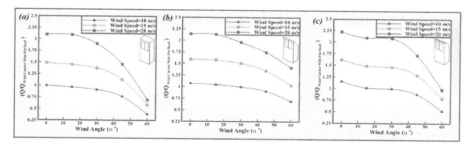

FIGURE 6.18 Changes in air flow rate due to wind speed and wind direction. (a) Wind catcher with flat roof. (b) Wind catcher with steep roof. (c) Wind catcher with curve roof [18].

FIGURE 6.19 External view of solar wind catcher [49].

dampers and, as the humidity increases, it decreases the temperature as shown in Figures 6.18 and 6.19. This solar wind catcher model works like a conventional wind catcher in normal conditions, but in sunny and high-temperature conditions, the fan should move and the damper control part must check the moisture content and all of this automatically gets triggered by sensors as embedded in the system as shown in Figure 6.20. The solar wind catcher is currently the best way to upgrade the traditional wind catchers [35].

The mono-drag wind catcher works in every condition: in the winter, in the evenings and on the weekends, as the building is empty. The wind catcher system does

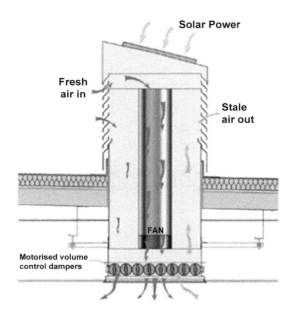

FIGURE 6.20 An example of modern wind catcher and its performance [49].

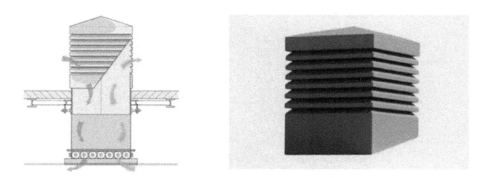

FIGURE 6.21 Mono-drag wind catchers [50].

not depend on pop-ups or sideway doors and allows the building to be completely secure. The system also works at night, uses the cooler air at night to drain heat from the building and cool the room and prepare it for the next day. The wind volume controls dampers that are located at the base of the system in the ceiling and controls the amount of air flow into the system. If the inlet temperature is less than 15°, the damper is automatically closed to prevent overheating, as shown in Figure 6.21 [50].

Elmualim et al. [43] in 2006 concluded that the introduction of heat source in the speed test room would increase wind flow, especially at low wind speeds. During the heat source test, the maximum internal temperature was 30°C–32°C, while the outside temperature was 20°C–22°C. At a wind speed of 1–3 m/s, the increase in flow

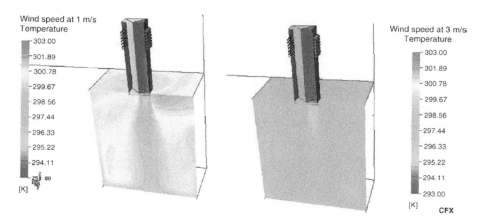

FIGURE 6.22 Temperature contours with heater source, with wind speeds of 1 and 3 m/s [43].

velocity varies from 7% to 54% respectively. In one case of flow velocity, the increase in flow velocity was 38% for an airspeed of 1 m/s and for airspeed of 3 m/s, the flow rate was reduced to 2%; they found that if the temperature difference between the outside and inside is 8°C; a heat source wind catcher can reduce the temperature to 6°C. In Figure 6.22, the results of the test are evident.

Elmualim and Sayigh [51], in their study, have used various predictive models to evaluate the performance of the wind catcher for the purpose of natural ventilation in the building. In these studies, wind tunnel tests smoke visualization, explicit, AIDA implicit model and CFX CFD code have been used. The wind catcher function depends considerably on the direction and speed of the wind catcher. Wind tunnel tests showed that the amount of ventilation increases with wind speed and decreases slightly by increasing the wind angle, in the range from 0° to 45°, especially for winds with a low velocity. The results obtained using Explicit, AIDA, and CFX code were related to experimental results for winds with a lower velocity, with an angle of 0°–15°, while CFX results showed a higher angle of 30°–45°. This difference can be attributed to the difficulty of accurately measuring the flow of air due to geometric and network complications.

6.10.2 SOME EXAMPLES OF MODERN WIND CATCHERS ALL OVER THE WORLD

6.10.2.1 Prince Nora University of Riyadh, Saudi Arabia

For the design of this university, a number of courtyards are included in various scales. In its largest scale, every four quadrants act as a common ground between two educational spaces, which together create a large public space. This yard is cooled with large passive wind catchers that absorb the breeze and lower the air flow to the surface [52]. It is shown in Figure 6.23.

6.10.2.2 Doha University of Qatar

This university is located in the Doha city of Qatar and is designed according to the conditions of Persian Gulf climate (Figure 6.24).

FIGURE 6.23   Pictures of the wind catchers at the University of Prince Nora, Riyadh, Saudi Arabia [52].

FIGURE 6.24   Wind catchers of the University of Doha, Qatar [53].

FIGURE 6.25   Images of the modern city of Masdar and its wind catchers in Abu Dhabi, UAE [54].

6.10.2.3   Masdar City

The city of Masdar, which is projected to be the world's first carbon-neutral city, has been engineered to be a cool island, although it is surrounded by desert climate. This city focuses on science and technology of advanced substitute energy, environment, technologies and sustainability through a variety of new sustainable modern fans, such as the use of wind catchers, located in the right place [52] (Figure 6.25).

6.10.2.4   Blue Water Shopping Center in England

The Blue Water Building has three right-wing shopping malls in direct line with reception halls and surplus services. These three shopping centers are located south, east, and west, with the courtyard in the middle. Shops on two

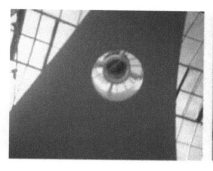

FIGURE 6.26 The wind catchers at the British Blue Water Shopping Center from inside and outside [54].

floors are interpreted in these three shopping centers. Shopping centers are ventilated by a combination of natural ventilation by wind catchers and air conditioning systems located on the ceiling of the building. Thirty-nine rotary wind catcher units are there with conical shape, distributed throughout the shopping centers, each shopping center including 13 wind catcher systems, as shown in Figure 6.26 [54].

6.11 INTRODUCTION TO THE APPLICATION OF GEO-ENVIRONMENTAL ENGINEERING AND SUSTAINABLE DEVELOPMENT KNOWLEDGE IN PROMOTING ENVIRONMENTAL QUALITY

In the 20th century, population growth, industrial development, and the lack of attention to the environment and the destruction of natural resources led groups of people and international organizations, such as the United Nations, to emphasize the importance of protecting the environment and preventing its destruction for future generations.

Measures such as the establishment of the United Nations Environment Program (UNEP) and also the United Nations Conference on Environment and Development (UNCED was a major United Nations conference held in Rio de Janeiro in 1992 which was called the "Earth Summit") were part of these activities to protect the environment, that in various countries was required to adopt strategies and take measures to protect the environment. After that, familiar phrases like "Thinking globally acting locally" and "We only have one Earth" and words such as "Sustainable Development", "Clean Air", "Safe Water", and "Recycling" in the literature of most countries became popular. Environmental science has become more used to prevent contamination of biological resources such as water, soil, and air. Also, environmental issues were addressed. The task of environmental engineering is to use the knowledge and ability of engineering to reduce the dangers that threaten human and its life on the planet.

The subject of this section is a combination of environmental engineering and geotechnical engineering. This combination in technical literature is considered in two cases:

1. Geo-environmental Engineering
2. Environmental Geotechnics

The first issue is a specialized branch of environmental engineering that deals with the pollution of soils and underground water tables. The second issue is a branch of geotechnical engineering that analyzes, designs and constructs structures that have environmental objectives.

The construction of wind catchers will be very effective in protecting the environment, so we consider these structures in the second part. In fact, by constructing these structures, we will directly contribute to environmental protection, as these structures will replace industrial cooling and air conditioning systems. In addition to the stunning architecture of the wind catchers and the beauty of urban space, we have saved the environment from the toxic and harmful gases that are produced by these devices. These are important factors that can cause irreparable damage to the environment. At the moment, the world is rapidly consuming and eliminating energy, building structures that are effective in saving and reducing energy consumption will be very important and valuable. These structures simply circulate air inside the structure solely because of their special design and this happens without any energy consumption. Therefore, we are entering a part of a sustainable development that makes the situation more suitable for us and future generations. In addition to preserving natural resources, the use of these structures in the economic sector will also be effective.

6.12 CONCLUSIONS

The aim of this research is to achieve a comprehensive view of natural ventilation by wind catcher. In this chapter, the types of wind catcher are introduced in a traditional and modern style. The mechanism of ventilation in the modern and traditional wind catchers and their history have been studied. Also, a comprehensive overview of researches in the field of improving wind catcher performance is presented in this study. By examining the type of materials used in Iranian wind catchers, it is concluded that not only they can be favorably considered in the field of domestic air quality and provide air conditioning and human comfort, but also have other green features such as the type of indigenous material, cost reduction and energy consumption. They also help in reducing the negative impacts on regional and global environment and have always been an integral part of the houses in the deserts of Iran. They used to be made with such materials that were most energy efficient.

An indigenous home is a smart home because it ensures a comfortable interior for a long time. These homes can inspire smart design in today's houses. Nowadays, the use of natural ventilation in houses, especially smart homes has been considered to improve indoor air quality. The use of artificial intelligence (AI) and the Internet of Things (IoT) in the design of wind catchers, as well as in controlling the location of the wind catcher and the amount of wind entry for natural ventilation of smart homes can be useful. Of course, resource constraints and concerns about reliability are the two major challenges of smart homes, and planning must be done to address potential problems and issues.

In terms of sustainability, the wind catcher has a great deal of coordination with its environment. It is environment friendly and does not have any negative effects on humans, while at the same time it minimizes energy consumption for cooling and air conditioning in the building.

REFERENCES

1. Commission, B., *World Commission on Environment and Development. Our Common Future.* 1987, Oxford, United Kingdom: Oxford University Press.
2. Council, U.G.B., *LEED Reference Guide for Green Building Design and Construction for the design, Construction and Major Renovations of Commercial and Institutional Building Including Core & Shell and K-12 School Projects* 2009 ed. 2009. U.S. Green Building Council. Washington, USA. Available from: https://www.usgbc.org/resources/-leed-reference-guide-green-building-design-and-construction-global-acps
3. Montazeri, H. and R. Azizian, Experimental study on natural ventilation performance of one-sided wind catcher. *Building and Environment*, 2008. **43**(12): pp. 2193–2202.
4. Alp, A.V., Vernacular climate control in desert architecture. *Energy and Buildings*, 1991. **16**(3): pp. 809–815.
5. Al Suliman, A., Wind catchers and sustainable architecture in the Arab world. *Civil and Environmental Research*, 2014. **6**: pp. 130–136.
6. Tavakolinia, F., *Wind-Chimney (Integrating the Principles of a Wind-Catcher and a Solar-Chimney to Provide Natural Ventilation), in Architecture.* 2011, California Polytechnic State University. San Luis Obispo
7. Wu, Y.C., A.S. Yang, L.Y. Tseng, and C.L. Liu. Myth of ecological architecture designs: Comparison between design concept and computational analysis results of natural-ventilation for Tjibaou Cultural Center in New Caledonia. *Energy and Buildings*, 2011. **43**(10): pp. 2788–2797.
8. Karava, P., T. Stathopoulos, and A.K. Athienitis, Wind-induced natural ventilation analysis. *Solar Energy*, 2007. **81**(1): pp. 20–30.
9. Saini, J., M. Dutta, and G. Marques, Indoor air quality prediction systems for smart environments: A systematic review. *Journal of Ambient Intelligence and Smart Environments* (Preprint), 2020. **12**(5): pp. 433–453. DOI: 10.3233/AIS-200574
10. Salman, M., S. Easterbrook, S. Sabie, and J. Abate, Sustainable and smart: Rethinking what a smart home is. In *ICT for Sustainability*. 2016 August (pp. 184–193), Atlantis Press. Paris. France. DOI: https://doi.org/10.2991/ict4s-16.2016.22
11. Yang, Y.K., M.Y. Kim, Y.W. Song, S.H. Choi, and J.C. Park, Windcatcher louvers to improve ventilation efficiency. *Energies*, 2020. **13**(17): p. 4459.
12. Khan, N., Y. Su, and S.B. Riffat, A review on wind driven ventilation techniques. *Energy and Buildings*, 2008. **40**(8): pp. 1586–1604.
13. Faggianelli, G.A., A. Brun, E. Wurtz, and M. Muselli. Natural cross ventilation in buildings on Mediterranean coastal zones. *Energy and Buildings*, 2014. **77**: pp. 206–218.
14. Bahadori, M.N., An improved design of wind towers for natural ventilation and passive cooling. *Solar Energy*, 1985. **35**(2): pp. 119–129.
15. Chiesa, G., M. Grosso, D. Pearlmutter, and S. Ray., Advances in adaptive comfort modelling and passive/hybrid cooling of buildings. *Energy and Buildings*, 2017. **148**: pp. 211–217.

16. Branch, C.T., Ancient Iran, the origin land of wind catcher in the world. *Research Journal of Environmental and Earth Sciences*, 2013. **5**(8): pp. 433–439.

17. Zarandi, M.M., Comparative analysis on architectural characters of Iranian wind catchers in hot arid (case study: Yazd & Bandar Lengeh). *International Journal of Advanced and Applied Sciences*, 2015. **2**(8): pp. 17–22.

18. Dehghan, A., M.K. Esfeh, and M.D. Manshadi, Natural ventilation characteristics of one-sided wind catchers: Experimental and analytical evaluation. *Energy and Buildings*, 2013. **61**: pp. 366–377.

19. Yaghoubi, M., A. Sabzevari, and A. Golneshan, Wind towers: Measurement and performance. *Solar Energy*, 1991. **47**(2): pp. 97–106.

20. Jafarian, S.M., S.M. Jaafarian, P. Haseli, and M. Taheri., Performance analysis of a passive cooling system using underground channel (Naghb). *Energy and Buildings*, 2010. **42**(5): pp. 559–562.

21. Pirnia, M.K., *The Islamic Architecture of Iran*. Memarian, G. ed., 2002. Soroush Danesh. Tehran, Iran.

22. Watson, D., *Climatatic Design: Energy Efficient Building Principles and Practices*. 1983. McGraw Hill Higher Education. New York. USA.

23. Zarandi, M.M., Analysis on Iranian wind catcher and its effect on natural ventilation as a solution towards sustainable architecture (case study: Yazd). *Engineering Technology*, 2009. **54**: pp. 574–579.

24. Mahyari, A., *Wind Catchers*. Unpublished Ph.D. thesis, Sydney University, Australia, 1997.

25. Fathy, H., Natural energy and vernacular architecture 1986/Hassan Fathy-Chicago: Published for the United Nations University by the University of Chicago Press. XXIII, 1986.

26. Bahadori, M.N., Passive cooling systems in Iranian architecture. *Scientific American*, 1978. **238**(2): pp. 144–155.

27. Saljoughinejad, S. and S.R. Sharifabad, Classification of climatic strategies, used in Iranian vernacular residences based on spatial constituent elements. *Building and Environment*, 2015. **92**: pp. 475–493.

28. Koch-Nielsen, H., *Stay Cool: A Design Guide for the Built Environment in Hot Climates*. Routledge; 2013 Jun 17.

29. *Ancient Wind Catchers in Hyderabad*. 2017 [cited 2017 9/9/2017 8:00 PM]; Available from https://www.insideflows.org/project/ancient-wind-catchers-in-hyderabad/

30. Hughes, B.R. and S.A. Ghani, A numerical investigation into the effect of windvent dampers on operating conditions. *Building and Environment*, 2009. **44**(2): pp. 237–248.

31. Tolba, M.M., Wind towers "Wind Catchers" a perfect example of sustainable architecture in Egypt. *International Journal of Current Engineering and Technology*, 2014. **4**(1), pp. 430–437.

32. A'zami, A. Badgir in traditional Iranian architecture. In *International Conference "Passive and Low Energy Cooling for the Built Environment"*, Santorini, Greece. 2005. pp. 1021–1026.

33. Mahmoudi M. *Wind Catcher: The Symbol of Iranian Architecture*. Tehran: yazda. 2007.

34. Mofidi, S., Analysis on typology and architecture of windcatcher and finding the best type. *Honar-Ha-Ye-Ziba Journal*, 2008. **36**: p. 29.

35. Jazayeri, E. and A. Gorginpour, Construction of windcatcher and necessity of enhancing the traditional windcatcher. *Magazine of Civil Engineering*, 2011. 6(24): pp. 56–60. DOI: 10.5862/MCE.24.8.

36. Ghaemmaghami, P.S. and M. Mahmoudi. Wind tower a natural cooling system in Iranian traditional architecture. In International Conference "Passive and Low Energy Cooling for the Built Environment 2005 May. Santorini, Greece. p. 73.

37. Ghobadian, V., *Climatic Analysis of the Traditional Iranian Buildings.* 1998, Tehran, Iran: Tehran University press.
38. Zamarshidi, H., *Constricting Building with Masonry Materials.* 2005, Zomorod Publication, Iran.
39. *Lecture 5a Passive Ventilation.* Spring 2012 [cited 2017 20/8 7:00 PM]; Available from http://isites.harvard.edu/
40. Oates, D., Innovations in mud-brick: Decorative and structural techniques in ancient Mesopotamia. *World Archaeology,* 1990. **21**(3): pp. 388–406.
41. Bater, M., H. Ahmadi, and R. Emadi, Investigation of silicates micronized additives effects on erosion under rainfall simulation and permeability of Kahgel traditional plaster. *JWSS-Isfahan University of Technology,* 2017. **21**(1): pp. 185–201.
42. Kim, J.J., and B. Rigdon B. *Sustainable Architecture Module: Introduction to Sustainable Design.* 1998, National Pollution Prevention Center for Higher Education. Ann Arbor. 1998.
43. Elmualim, A.A., Effect of damper and heat source on wind catcher natural ventilation performance. *Energy and Buildings,* 2006. **38**(8): pp. 939–948.
44. Chaudhry, H.N., Achieving sustainable buildings: The role of heating, ventilation and air-conditioning. *Sustainable Buildings,* 2016. **1**: p. 1.
45. Abdallah, A.S.H., Experimental study of passive air condition system integrated into a single room in Assiut, Egypt. *Energy and Buildings,* 2017. **153**: pp. 564–570.
46. Gharakhani, A., E. Sediadi, M. Roshan, and H. Bagheri Sabzevar, Experimental study on performance of wind catcher in tropical climate. *ARPN Journal of Engineering and Applied Science,* 2017. **12**(8): pp. 2551–2555.
47. Badran, A.A., Performance of cool towers under various climates in Jordan. *Energy and Buildings,* 2003. **35**(10): pp. 1031–1035.
48. Calautit, J.K., B.R. Hughes, and D.S. Nasir, Climatic analysis of a passive cooling technology for the built environment in hot countries. *Applied Energy,* 2017. **186**: pp. 321–335.
49. *Sola Boost Windcatcher / Monodraught co.p.3* 2017 [cited 2017 10/9/2017 9:00 PM]; Available from https://www.monodraught.com/products/natural-ventilation/sola-boost-classic
50. *Classic Windcatcher.* 2017 [cited 2017]; Available from https://www.monodraught.com/products/natural-ventilation/windcatcher-classic
51. Elmualim, A.A., and A.A. Sayigh, *Modeling of a Windcatcher for Natural Ventilation.* InWorld Renewable Energy Congress VIII (WREC 2004) 2004 Aug 29 (Vol. 29). National Renewable Energy Laboratory Denver, CO.
52. Elmeligy, D.A.. Innovative sustainable technologies in heritage revival. *New Arch-International Journal of Contemporary Architecture.* 2014. **1**(2): pp. 101–111.
53. 2017 [cited 2017 8/30 6:28 PM]; Available from http://www.qu.edu.qa
54. Elmualim, A.A., Failure of a control strategy for a hybrid air-conditioning and wind catchers/towers system at Bluewater shopping malls in Kent, UK. *Facilities,* 2006. **24**(11/12): pp. 399–411.

7 Digital Twin and the Detection and Location of DoS Attacks to Secure Cyber-Physical UAS

Gita Donkal
Chandigarh University

Anjali Donkal
National Institute of Technology

CONTENTS

DOI: 10.1201/9781003132868-7

7.1   INTRODUCTION

Michael Grieves was the first person to introduce digital twin concept in 2002 at the University of Michigan [1]. Digital twin technology (DTT) is also known as computational mega model, avatar, device shadow, synchronized virtual prototype or a mirrored system. Not only in designing and operating cyber-physical intelligent systems, does DTT play its significant transformative role, but it also has the potential to upgrade the modularity of multi-disciplinary systems for tackling the rudimentary hurdles which are not addressed by the present modeling techniques [2]. With the emergence of Internet of Things (IoT), digital twin technology has also evolved as a dynamic virtual model of a process, system or service. Digital twin has the potential to incorporate sensor and contextual data from physical processes (or systems) into its computational model that is capable of solving circumvent problems and developing informed technology. Nowadays, one of the most integral components of the cyber-physical systems is digital twin technology where systems and operations are connected [3]. From the researchers of NASA and US air force, it has become evident that digital twin paradigm allows not only to fully comprehend the anomalous and degradation events but also make the prediction of previously unknown events which determines the success and failure of mission. Digital twin technology is aimed at finding solutions for the shortcomings of current paradigms and practices for certification, sustainment and fleet management [4].

It is evident that with new technologies come with several challenges and cyber-attacks are one of them. Out of all the multitude of vital concerns in the context of security, to DTT is the most immediate representation of an actual system and if the twin is procured by a hacker, then undoubtedly, it will serve as a prototype to the original system which will be identifying components, their interfaces and behaviors. Indeed, this assumption can be made that before any attack on physical system, a complete script is mapped out by the hacker, making use of vulnerable DTT, permitting penetration of the real system with minimal disruption or detection. Performing pen-testing on the interfaces of physical systems using digital twins, authorizes the hacker to fine-tune their attacking mechanisms [5].

DTT is a procedure that allows autonomous objects such as products, machines, etc. to converge the present state of their procedures and behaviors in context to the ecosystem of the actual world. This way, manufactured commodities could increase the employment of connected cyber-physical data to form smart commodities that assimilate capabilities such as self-management, entirely based on connectivity and computational technologies [6].

Attacks, including Denial of Service and spoofing, have revealed their true potential over the past few years and these cyber-attacks are growing more powerful with the passage of time. These attacks could be ignored for a while when launched on civil applications, but when they target military of a nation, indeed it is not possible to ignore them anymore. In the 21st century when computers have proven their noteworthy progress in various fields and domains, including threatening cyber-crime world, securing physical aerial assets and patrolling the perimeters would not be sufficient to safeguard a nation.

The new kind of attack is cyber-attack that can be launched massively from anywhere on this earth with a network connection. DoS attacks are one of a kind, as they are capable of leveraging all computational resources and prohibiting the legitimate user to utilize the resources. Once the genuine user is denied of a service, it can be create havoc, since attacker will have full authorization and leverage to operate aerial vehicles. They can navigate them off the given itinerary or trajectory and it can also get worse than that [7]. Therefore, there is an urgent need to come up with novel ideas, critical thinking and improving existing ideas. Unmanned aircrafts are now being equipped with more cutting-edge functionalities and techniques, including improvement in the ISR (Intelligence, Surveillance, and Reconnaissance). The idea of detecting, locating and mitigating a DoS attack itself can avoid multiple forms of attacks including impersonation, signal jamming, GPS and GNSS spoofing attacks and many more.

UAVs functionality can be severely compromised on the launch of DoS attack as it can disturb and incapacitate network availability, GPS navigation system, video streaming operations display false negative data and much more [8]. Figure 7.1 depicts various wireless attacks that can be launched on UAVs.

Cyberspace is heavily used to support shaping and influence operations. Six major categories include insiders, hackers, corporate spies, terrorists, criminals and nation-states. Security threats are often classified on the basis of the nature of attacker and its mission. Some of the actions taken by a legitimate user to prevent their assets from any harm caused by an adversary include:

- Deny that prevents adversary from using resources,
- Degrade that includes denying access to a target to a level, represented as a percentage of capacity,
- Disrupt, which is also a very specific case of degradation, where the degradation level selected is 100%,
- Destruction is the permanent denial of access to operation of a target and manipulate controls or changes the attacker's systems, information, and networks [9,10].The compromised functionality of ECU could devastatingly change functionality of an autonomous vehicle, just because of the disrupting DoS attack. Therefore, it is significant to run comprehensive analysis that can predict how and to what extent ECUs can be effected. We have been using GPS for a very long time, however, we still don't know the potential of networked GPS that could be used to track down our enemy (attacker). Researches using a networked GPS to counter present and future attacks, especially DoS attacks have not been carried out much and need more attention from researchers [11].

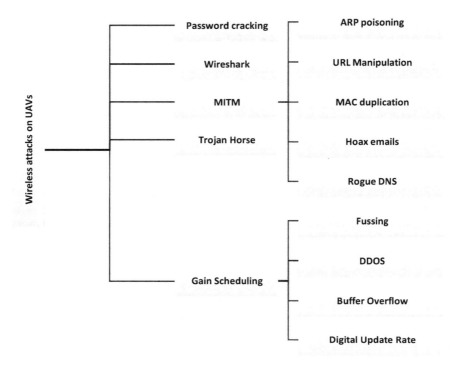

FIGURE 7.1   Multifarious wireless attacks on UAVS.

Very much like the ransom-ware WannaCry that was launched by a youngster who just found a susceptible spot in Microsoft's Windows OS and was capable of affecting a hundred nations at a time and able to shut down all network connected computers for some time by locking out the legitimate user and taking control of their valuable files [12], I personally do not see any reason why this can't happen to UAVs. Indeed, there can be potential implications of an automated attack, executed on a large-scale without expert knowledge. Henceforth, there is a need to hire cyber security experts that not just think speculatively but also are capable of spotting probable vulnerable spots with high-tech tools, codes and algorithms.

Additionally, autonomous aerial vehicles these days can be trained to be smart enough to detect and prevent adverse autonomous activities, caused by DoS attacks. This gap needs more investigation and consideration since it revolves around DoS attacks which are capable of shutting down all sorts of communication and control systems [13].

Especially, the DTT can be used for:

- **System model validation with real world data**: In order to validate the DTT model and make the evaluations and predictions, assimilation of operational ecosystem data and the interlinkage of the system with that ecosystem into the DTT becomes essential.

- **Facilitating decision support and notifying the users**: Following the assimilation of operational, maintenance and health data, the DTT can be implemented in a what-if analysis mode for producing information regarding customized decision support and alerts to users of the physical system.
- **Prediction of modifications in the physical system**: Operations' optimization, contingency planning enhancement and system performance prediction are facilitated by an analysis based on simulation of operational and maintenance data from the physical twin. For the prediction of modifications in the physical systems as well as its parameters, DTT can be incorporated for dealing with contingencies [14].

The aim is also to study more about GPS and GNSS systems since we will be using them to detect and mitigate spoofed attacks. Additionally, the rudimentary purpose is to get familiar with potential offensive and defensive cyber space operations and to learn about the responsibilities of military of a nation, whenever a cyber-attack is launched on valuable asset. Last but not the least, the major objective of this research is to identify the potential of DoS attacks, its multifarious forms, detection mechanisms and how to mitigate incoming spoofing attacks taking either proactive or reactive measure.

Figure 7.2 shows a simplified diagram of GPS spoofing attack, where the attacker has overpowered an authentic GPS signal and represents three different scenarios. Considering first and foremost scenario, the attacker is launching a form of DoS

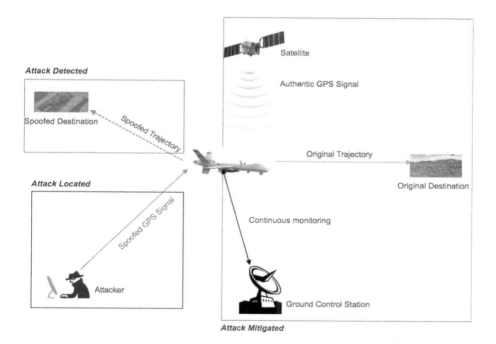

FIGURE 7.2 Attacker compromising a UAV using spoofed GPS signal.

attack, which is a GPS spoofing attack. He has successfully spoofed the incoming original signal from a satellite GPS and now our task is to detect this form of attack.

In second scenario, the adversary has succeeded in spoofing the trajectory of the UAV that was on its way to the original trajectory but instead of reaching to its legitimate and un-spoofed destination, it has reached to a spoofed destination, where this UAV can be hijacked and can be used for reconnaissance by the adversary. That leaves us with no choice but to locate it since without knowing its location, we can't proceed to take any defensive action. Eventually, in third scenario, this attack needs to be dealt with a strong mitigation technique, so that the UAV can be put back on its original and non-malicious trajectory [15,16].

Nowadays, cyber defense philosophies make limited use of military strategies and tactics. Defense commanders realize the value of deception and maneuver, and defense-in-depth philosophy, the strategies are static and their tactics are reactive [17]. Nevertheless, infrastructure and aerial vehicles could be vulnerable and simultaneously any security mechanism could fail to operate or could also become susceptible to current situation, then what should be our back up or recovery plan? [18]. However, the true potential of an attacker is always unknown as there are anonymous ways to launch DoS attacks and to counter reply these impeccable attacks, we never know how prepared we are. There is always a fear of being victimized and possibilities of taking large-scale reactive mandatory actions to defend cyber-physical unmanned aerial crafts. This needs to be explored [15].

Since in DoS attacks, the IP address looks quite original therefore it is not a cakewalk to identify the spoofed IP address that arises the need to develop more robust systems and using this research our objective would be developing an implementable model that can easily distinguish between legitimate and malicious IP address.

Rest of the paper is organized as follows. Section 7.2 discusses some of the terminologies associated with techniques used in UAVs for the detection of attacks and a complete surveillance on them. Section 7.3 assimilates the knowledge of various possible ways of launching DoS attacks on a UAV along with a table that highlights some methods of safeguarding UAVs against spoofing attacks. Section 7.4 entails some experimentation and results of different DoS attacks on UAVs. Finally, Section 7.5 concludes the paper followed by Section 7.6 that throws light on some of the scopes for future.

7.2   RELATED TERMINOLOGIES

In this section, an assimilative review of literature including various aspects to security of UAVs, contribution of various navigational and surveillance systems to work in collaboration with each other is provided. Multifarious reasons and strategies of attacks being launched by adversaries is explained more in detail, especially DoS attacks. Factors that motivated the approach taken in this research are also discussed.

7.2.1   DIGITAL TWIN AND UNMANNED AERIAL VEHICLES (UAV)

A Digital Twin is considered as a multiscale, incorporated multi-physics, probabilistic simulation of an as-built system that makes use of the best available sensor updates, physical models, fleet history, etc. for mirroring the life of its corresponding flying twin. Additionally, the DTT integrates data of sensors taken from the vehicles' on-board system, Integrated Vehicle Health Management (IVHM), maintenance history and all the accessible fleet and historical data procured using data and text mining. The DTT can predict about the system response against the safety critical episodes and unveil the formerly concealed issues before they turn out to be critical after making comparisons of predicted responses and actual responses [4]. Being so ultra-realistic, DTT takes into consideration one or more integral and interconnected vehicle systems including propulsion, avionics, energy storage, thermal protection and life support. Eventually, the alleviation of damage or its degradation for the systems on-board the digital twin can be achieved by setting self-healing technology in motion, or through recommendation of changes in mission profile that will help in procuring lesser loads, consequently achieving both the probability of mission success and life span [4].

Cyber-physical systems (CPS) are new-generation systems incorporated with physical and computational capabilities. UAVs are CPS as they are reliant on the interaction between physical and computational elements of the system [7]. UAVs are also known as drone, remotely flown target aircraft and remotely piloted aerial vehicle (RPAV). UAVs are categorized depending on the purpose they are used for that indulges target and decoy, reconnaissance, combat, logistics, research and development, and civil and commercial. Other than that, these days UAVs are being used heavily for military, commercials, surveillance, recreational and agricultural purposes, aerial photography, smuggling, product deliveries and much more. UAVs cover everything from aerodynamics of drone, materials in the manufacture of physical UAV, to circuit boards, and chipset, a software- which is also the brain of a drone [19,20]. Drones are made up of light composite material to reduce weight and to enhance maneuverability so that they can cruise at high altitudes. Also, they are equipped with infrared cameras, GPS and LASER (Light Amplification by Simulated Emission of Radiation), especially UAVs used for military. These drones are controlled remotely by ground cockpit [21].

Figure 7.3 depicts the modelling framework of a cyber-physical system which helps in understanding the layout of interactions taking place between its two

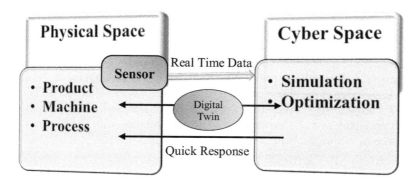

FIGURE 7.3   Modelling framework of the cyber-physical system [6].

integral components, including physical space and cyber space. Furthermore, the interaction of subcomponents of physical space leads to product, machine and process, whereas for cyber space, it includes simulation and optimization that are well represented for real-time data. It also shows the use of DTT that facilitates highly accurate and quicker responses in comparison to the current technology [6].

UAVs have the potential to operate without an international pilot and can be flown by a remote pilot on a preset flight path. Depending on the kind of sensors adhered to them, UAVs have a range of potential applications including disease detection, weed detection, monitoring responses to fertilizer spraying aerially [8]. UAVs are embedded with cameras that are capable of collecting high-resolution images and sensors that can easily compile huge amount of data in order to assist with monitoring and decision making in aerial battlefield. UAVs are also capable of making fast decisions and also have the potential to collect data with very high spatial resolutions. UAVs also have the potential to provide immediate visual information regarding large areas that can help in speeding up the decision making process [22]. However, UAVs still have some limitations for working in battle field, including high initial costs, sensor capabilities, reliability and lack of standardized procedure to perform computations on large chunks of data [23].

7.2.2 Challenges with Current Technology/UAVs

- Record low false negatives and false positives.
- Adaptation of novel circumstances (to generalize), including degradation or failures of subsystems.
- Learning from past experiences and effective integration.
- Promptness and availability of system data with enough on-board processing resources for analyzing and archiving present/past data trends respectively.
- Anticipation of futuristic events and states such as predictive modelling of imminent faults and other useful life bottomed on present/historical data trends [24–26].

7.2.3 Pros of Digital Twin

- DTT is a specimen that considers performance, structure, mission-specific characteristics including malfunctions confronted, miles covered, and restoration and maintenance history of the physical twin.
- It assists in determining preventive maintenance schedule set up on the basis of knowledge of the observed behavior and history of systems' maintenance.
- DTT allows a user to understand the way physical twin performs in the actual world, and the expectancy of its performance with prompt maintenance in the future.
- Traceability is ensured amongst different phases of life cycle via connectivity, facilitated by digital thread.
- Refined conjectures can be provided along with predictive analytical data gathered from tangible systems and they can be further assimilated in the

DTT to analyze in parallel to other information sources, for making predictions about performance of system in the future.

- It is also capable of troubleshooting malfunctions confronted by remote equipment along with their maintenance.
- Physical systems' data and IoT data is combined using DTT that facilitates service optimization, manufacturing processes and identification needed for design enhancements.
- It assimilates the operational and maintenance data from the tangible systems into its models and simulations [14].

7.2.4 DETECTION SYSTEM FOR UAVs

To perform communication, navigation, and surveillance, we need effective and highly proficient detection systems. Upcoming section discusses some of the most valuable and integral detection systems.

7.2.4.1 Satellites

Satellites are majorly used for navigational purposes to locate exact longitudes and latitudes. GNSS comprises a constellation of orbiting satellites in conjunction with a network of ground station. The signal information is converted by a receiver into position, velocity and time. This information can be used by any receiver to determine the exact position of a satellite. For SAR (Search and Rescue) missions, high-end GNSSs are taken into consideration that incorporates satellite constellation providing a much better coordinated coverage and synchronization with each other [12,27]. Radio Detection and Ranging (RADAR) technology in drone will signal that enough drone GNSS satellites have been detected and drone is ready to fly. It also displays current position and location of drone in relation to pilot on remote controller display and also records home point for return to home (R2H) safety feature [28,29].

To extend, there are three types of return to home techniques that include pilot initiated R2H by pressing button on remote controller or in an application. Second, low battery level, where UAV will fly back automatically back to home and third, laws of transmission between UAV and remote controller ensure that the UAV flies back automatically to its home point. To avoid obstacles, UAV makes the use of vision sensor, ultrasonic, infrared, Light Detection and Ranging (LIDAR), time of flight and monocular vision, also including GPS, speed of movement, increased level of computational functionalities and much more. To be more cautious, one can always use multiple GPS satellites to provide better coverage, generation of rogue signal to mislead or block GPS device and military GPS systems can rely on encrypted signals [30,31].

7.2.4.2 Kalman Filter Algorithm and RADAR

It is named after Rudolph E. Kalman and used for optimal estimation of location, speed and direction. Its common application includes guidance and navigation systems, computer vision systems and signal processing. It is used to estimate the trajectories of manned spacecraft to the moon and back. With the help of this algorithm, the variables of interest can be measured indirectly. It helps in increasing efficiency and sustainability of many of our current practices. Automated cars that enhance its traffic flow by reducing pollution, allow drivers to work or relax while in transit. Mass customization of product that closely match consumer preferences and reduces waste during production includes Tele-care alarm systems and CPS treatment tools. It is used for drones and military robots to defend nation while minimizing risk for military personnel [32,33].

To constantly track and acquire trajectory information can be called as state estimation problem. For filtering stochastic measurement errors in linear radar systems, Kalman filter has been adopted, although for practical applications, non-linear systems are more common [34]. To exemplify, for applications involving tracking down targets, space rectangular coordinates are used to set up motion models, whereas 3-D spherical coordinates are used to set up measurement models [35]. It is crucial to transform coordinates while processing the data, which is also a very typical, non-linear problem for which Kalman filter would be inappropriate to use to address this issue.

In order to filter random errors for motion models, Kalman filter and its variants can be assimilated. Some of the previous works of Parkinson et al. [25] presented an ample number of tracking objects with mobile radars. The same approach can also be used to track UAVs with mobile radars. Radars on UAVs and vehicles are two such types, which are depicted in Figure 7.4 [28].

Radio Detection and Ranging (RADAR/Radar) is one of the most integral detection systems of UAVs that are used for surveillance purposes. In order to improve the operability of UAVs in terms of obstacles detection and other traffic, a combination of non-cooperative sensors like active/passive forward-looking sensors, traffic collision avoidance system (TCAS) and most importantly RADAR, as a secondary

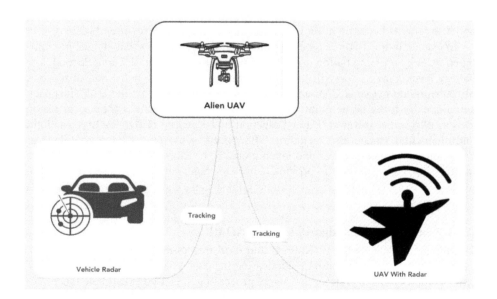

FIGURE 7.4   Tracking UAVs with mobile RADARS [55].

surveillance system is used [36]. Radar is also capable of performing multimode operation including range detection, Doppler sensing, SAR mapping and can also help in mitigating air collisions. Additionally, Radar has the potential to work in all weather conditions including fog, rain, and snow covering a much wider detection area than the field of view of a human pilot in an aircraft [37].

7.3 TAXONOMY FOR CYBER-ATTACKS ON UAVS/DTT

To generalize existing security threats, four major factors that these attacks can affect include confidentiality, integrity, availability, and authenticity. Confidentiality ensures reading the content of a secret message by only authorized entities. Integrity checks for the undetected alteration of the content of a message when it is being transmitted via network. Availability is the kind of security mechanism that is accountable for ensuring the availability of resources whenever they are requested by a legitimate user. Authentication is the mandatory duty of a receiver to confirm that the data received is from the correct and un-spoofed sender.

Assume a scenario, where a hacker successfully exploits the susceptibilities in a DTT, then the hacker will leverage this potential and uncover the organization to the core system attacks, since these DTT based systems are directly called by the twin and this threat will just unlatch the gates of backend/core systems to an attacker. On the contrary, tangible systems are frequently equipped with upgraded security features including micro-controllers that may not be facilitated on the platform running the twin. With mere analysis of code, a hacker can instantly locate the API calls needed for these backend systems and feasibly disclose the credentials for twins' access through authentication functions [38].

Code analysis, popularly known as reverse engineering is one of the most probable attacks on all platforms. Software designed for Linux OS or Windows can be conveniently reverse-engineered for two foremost reasons, firstly, they are widespread and secondly, the cost of reverse engineering tools is low. After procuring these access points, twin's behavior can be easily spoofed or even the tangible system can be accessed, conclusively facilitating access of system-wide data. The associated parties must acknowledge some foundational guidelines while designing and implementing digital twin for minimizing the risks involved in the development and functioning of a digital twin [39,40]. The numerous security artefacts can be inferred from digital twin including all the measures of physical protection, the equipment for access control, and the communication networks and its equipment for wireless communication. However, even after the inclusion of network equipment details in digital twin, the information required for cybersecurity analysis might not be available, for which these details should be added before extraction of data that is required for cybersecurity analysis [2].

Attacks are categorized in two types that can threaten entire cyber-physical model from its infrastructure to its software [9,10]

- Active attacks
- Passive attacks

7.3.1 ACTIVE ATTACKS

They target integrity, authenticity and availability of data that can be done using modification, fabrication and interruption respectively. Modification in data compromises integrity of data that can be done by fraudulently forging information. Fabrication affects authenticity aspect and can be executed, using counterfeited messages, UAV spoofing, GPS spoofing and much more [41]. Last but not the least comes interruption that is one of the most dangerous and disrupting form of attacks. It can be accomplished using DoS attack, routing attack, jamming of channels and various other means. Just the moment when we think that password cannot be hacked, there comes various means to crack these unpredictable letters [42]. There are three types of password attacks including dictionary attack, rainbow attack, and brute force attack. In future, when it wouldn't be a bottleneck to check so many possible combinations of alphabets, numeric keys and special characters with no computational complexities for instance using quantum computers, then it would be easy to break encryption using brute force attack. Tools or parameters that can be considered for detection of cyber-attacks on UAVs include Active Radar Surveillance that can use the rebound echoes on unmanned aerial ships to locate them and estimate their distance, approaching speed, penetration vector, short-term trajectory and size.

Comparatively, active eavesdropping is more dangerous than passive eavesdropping, for which there are two main reasons. Primarily, active eavesdropping aims to attack the main channel by degrading the channel capacity. As in Figure 7.5, to exemplify, an active eavesdropper is transmitting the jamming signals to the legitimate receiver in order to degrade the main channel's capacity. On the contrary, an active eavesdropper may also aim to improve the capacity of the eavesdropping channel.

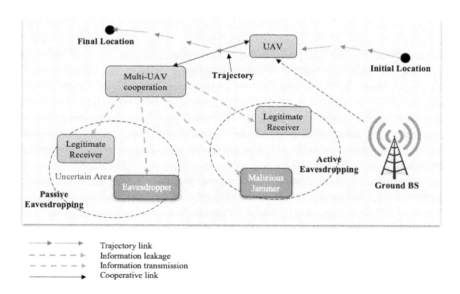

FIGURE 7.5 Illustration of Active and Passive eavesdropping leading to attacks like GPS spoofing [52].

Hence, active eavesdropping attacks are more detrimental than passive eavesdropping attacks. Especially, the improvements can be made in the eavesdropping channel's capacity by LoS channel characteristics when the unmanned aerial ship is a legitimate transmitter [40].

- **Active measures**: Active countermeasures can be brought into consideration here including missiles, cyber rifles or GTA (Ground-to-Air Defense) that has the ability to incapacitate the potential threat. Although they do come with some defensive issues such as less efficiency of GTA against cyber threats in urban zones where lives are at stake or concurrent attacks on multiple fronts [43].

7.3.2 PASSIVE ATTACKS

They target confidentiality of data using interception techniques. Interception can be carried out using traffic analysis, eavesdropping, launching viruses, malware, trojans, and so on. Eavesdropping can be executed using man-in-the-middle attack where the interception of data being transmitted between the transmitter and receiver nodes is done by an illegitimate, unauthenticated person and the attacker is illegally permissible to gather critical information of control systems, software, data and much more that can be used for reconnaissance purpose [44,45]. In traffic analysis scenario, an attacker can locate critical nodes of a wireless sensor network (WSN) in a UAV and can cause a huge disruption and devastation including forged navigational information [46]. Infecting valuable systems and software of a CPS with viruses like trojans and malwares, it can affect the overall functioning of CPS [9]. Heightened risk of a cyber security attack increases connectivity, automation and shared information among control systems and network of workstations leading to severe consequences.

- **Full position information of eavesdroppers**: A synthetic aperture radar equipped on the UAV can help detecting and tracking the positions of potential external eavesdroppers. If the eavesdropper stays stationary, obtaining the complete positions of these external eavesdroppers is possible at the UAV. Therefore, security concerns are likely to be addressed in such cases through proper resource allocation design taking advantages of the UAV's flexibility. However, a very high hardware cost is needed to obtain information regarding position with such precisions. Moreover, the additional load on the UAVs may increase its energy consumption [40].
- **Passive measures**: With a slight advantage over active measures, passive countermeasures are designed to safeguard the UAVs indirectly that includes the physical protections around target, some decoys, organized roadblocks, sensors jamming, and total or partial cyber-spoof of signals for GPS.

Similar to the tools for active attack detection, we also have tools for passive monitoring that can cover electromagnetic spectrum on a common communication channel through visible, thermal infrared, and radio waves [45].

7.3.3 MAN-IN-THE-MIDDLE ATTACK

To impersonate a legitimate user or operator in a UAV seems quite difficult, however an experiment was performed by Rodday et al., in which he launched an MITM attack on a UAV in one of his studies in 2016 that proved otherwise. If there is any loophole in the communication system of a drone, such as radio link in this case, it can be hacked in no time. There are some parameters associated with connection, including the DH and DL which are destination high and destination low addresses respectively. In order to access these addresses, one has to send broadcast packets confined to the radio network. Once UAV acknowledges these packets, the hacker can easily hack the link changing DH and DL, resulting in assimilation of reverse-engineering of the flight computer and UAV planning software that allows adversaries to modify and inject packets to interact with the drone's computer [47].

To gain full control of the UAV, this attack can be brought into consideration. This attack violates authentication protocols and standards that allows secure communication between the operator (user) and the UAV. In order to listen to the ongoing communication over a network that includes flights, Wi-Fi Pineapple Nano can be used [48]. This device scans for all the access points and imitates these access points that leads to connectivity between controller and Pineapple Nano. To avoid such type of eavesdropping, strong encryption techniques can be enforced, as there still exists some applications like Parrot AR Free-flight that directly sends data over HTTP that are highly vulnerable to MITM attacks [48].

7.3.4 DENIAL OF SERVICE (DOS)

This attack can be performed on a single network mode, V2V and V2I. DoS attack is capable of causing widespread disruption. It can disrupt traffic flow, prevent important communications and cause vehicle's collisions. DoS attack majorly affects internet-dependent and networked infrastructure components [49]. This is the kind of attack that is capable of preventing users to access a system for the duration and also it is more difficult to detect and defend against. DoS attack tools including Hping 3, Netwox, and LOIC (Low Orbit Ion Cannon) can be used to perform mid-air attacks [49]. In some scenarios, DoS attacks have the potential to crash a UAV.

Since DoS attack targets communication layers, it is very much possible to target any of the communication layers including physical link and network layer. Attacks that are more sophisticated, prevalent and harder to detect are directed usually at network layer, especially application layer [44] as shown in Figure 7.6. There are many ways to launch DoS attacks that can be classified as follows:

- **Jamming attack**: Jamming attacks are launched with the objective to interfere with legitimate wireless signals by emitting jamming signals. There are various techniques available to jam signals that can interfere with communication between two nodes affecting transmission of data, thereby bringing down the overall performance of Wireless Sensor Networks (WSN) [50].
- **IP spoofing attack**: A spoofed IP address can lead to discrepancy during navigation, and if the spoofed IP packets keep coming in the network,

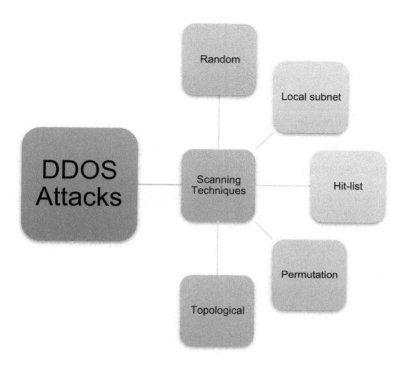

FIGURE 7.6   Different scanning techniques for DDoS attack.

then this discrepancy can escalate as well. In this attack, the attacker illicitly impersonates another IP address or machine by manipulating the IP packets. The IP packet can be forged with the spoofed IP address including source and destination address, checksum and much more [15].

- **Vampire attack**: Using some protocol parameters such as source routing, connection status and distance vector, vampire attack can be launched. Also vampire attacks can easily drain battery of network nodes. Adversary can create messages causing resources' depletion. Comparative to genuine code, an illegitimate node generates counterfeit messages that will not follow the optimal path towards the destination node which leads to unnecessary energy consumption [50].
- **TCP/SYN flooding attack**: In this attack, the attacker sends synchronization messages known as "SYN" packets in bulk to the victim. In response, victim sends "SYN-ACK" packet and waits for acknowledgement or "ACK" packet that attacker never sends back, leading to the buffer overflow condition and thus, the victim is not able to accept future connections [51].
- **Path-based DoS attack**: This attack injects fake and replayed packets in the network which in turn are forwarded by receiving node on the base station, thus wasting network bandwidth [16].
- **Wormhole attack**: To accomplish this attack, insertion of two nodes is required, where these two nodes will be connected by a strong connection

similar to a wired connection. To determine the shortest path, number of hops is calculated by the routing protocol. In this attack, the two rogue nodes identify a distant location with a single jump that can misguide other nodes on original distances between two nodes. This forces adjacent nodes to use rogue nodes for routing packets [50].

- **Black hole attack**: An illegitimate node is inserted in the network that modifies the routing table, forcing the maximum of adjacent nodes to route the information through it. Eventually, the rogue node acting as a black hole in the space that destroys entire information being transmitted [50].
- **Gray hole attack**: Gray hole attack is an improved variant of black hole attack. However, it retransmits a part of data including routing information and destroys valuable information, thereby making it more difficult to detect [50].

However, reinforcing existing systems and networks against DoS attacks can be a proactive measure to safeguard critical assets of aerial defense forces. Reacting appropriately to counter present and future attack trends falls in the category of reactive measure, which should be taken when a DoS attack is launched. Detection of attacks especially on navigation subsystems such as triggering of hovering or return-to-base mechanisms make foundational ground for reactive countermeasures.

Solution in abundance have been suggested by various researchers to cope up with DoS attack in a much strategic and reactive manner. To mitigate DoS attacks, upstream filtering can be used to relieve pressure on subsequent infrastructure. Intrusion Prevention Systems (IPS) can be deployed to automatically stop intrusion attempts upon their detection. Network communications that were identified in previous attack analysis can be ignored by using black holing of malicious traffic [52]. These days, data analysis has the capability to reveal hidden patterns, market trends, uncertain correlations and other such advantageous information that assist big organizations and industries to establish better cognizant commercial decisions [53].

Additionally, to avoid IP spoofing and DNS spoofing, a short-term mitigation approach can be adopted in which domain names can be redirected so as to alleviate attack impacts by altering or removing IP address, resolved by domain name. Routers and all network edge devices need to be configured according to the best practices along with system hardening that can be done by system administrator. Fusion of inertial measurement unit (IMU) sensor and GNSS systems with other sensors can be adopted as a mitigating methodology for spoofing attacks [54].

Table 7.1 shows different mitigation techniques adopted by various authors that are used to prevent spoofing attacks in a cyberspace.

Communicated breaking, platooning, traffic information system and local infrastructure network protocol attacks can be used to identify potential exploitation mechanisms; Communication protocols are analyzed. Moreover, attacks can be mounted and detected against CAN and Flex Ray protocols. One solution is to analyze data carefully and careful protocol analysis is required. With the number of Connected Autonomous Vehicles increasing on the road, the number of protocols to be examined also escalate with that [55].

TABLE 7.1

Different Mitigation Techniques to Prevent Cyberspace Spoofing Attacks

Authors	Methodology	Advantages	Drawbacks/Future Scope
Javaid et al. [55]	Receiver Autonomous Integrity Monitoring (RAIM) algorithm is used that determines integrity of GNSS solution. RAIM algorithm makes a comparison among pseudo-range measurements ensuring they all are consistent. RAIM exercises fault detection mechanism on computed set of navigational solutions isolating faulty satellites and providing mitigation level computation.	Integrity of position calculation using GNSS can be determined successfully. Once packets are received, RAIM is capable of detecting any attacking host.	For future scopes, further improvement of RAIM algorithm with quicker detection and correction can be considered.
Tanil et al. [44]	Use of aircraft autopilot response to deceptive trajectory to locate or detect GNSS spoofing attack. A tightly coupled INS-GNSS integration in Kalman filter is utilized to monitor spoofing attacks. It also investigates the impacts on spoofing attack detection due to aircrafts' dynamic response to control actions that are actuated by pilot/autopilot.	Even in worst case scenarios, spoofing attacks can be directly detected by the aircraft autopilot that provides a response to faulted GNSS signals.	
Mukherjee et al. [54]	It demonstrated three particular DoS attacks that uses J1939 data-link layer request and connection management protocols. Program ECU to drop incoming packets' request, lest already responded to the request from the same source address in a fixed interval of time.	This technique is useful to prevent overloading scenarios.	It can lead to exhaustion of resources as ECU will have to maintain state information.
He et al. [37]	A combination of visual sensor and IMU as a fusion of information in order to solve GPS spoofing issues.	Efficient use of UAV's own sensors with no auxiliary equipment requirement. Light weighted fusion algorithm. This approach can withstand sophisticated GPS spoofing attacks with acquisition of real-time flight speed.	Error accumulation during integration process. Gradual increase in cumulative error.

7.3.5 GPS Spoofing

GPS spoofing mainly targets a cyber-physical system that relies excessively on GPS for positioning and navigation that apparently shows the greatness of risk to an unmanned aerial system. GPS is responsible for providing navigation in endless sky as it provides exact coordinates, longitudes and latitudes to a UAV for navigation [56].

However, attackers are capable of spoofing GPS signals by transmitting fake or spoofed coordinates to the flight controller and taking over the full control of UAS. In order to spoof GPS signals, GPS has to be synchronized with the attackers' signals that comprises the spoofed information regarding almanac data. Instead of relying on GPS, DJI Phantom drones are utilizing inertial measurement unit (IMU) and they are still being targeted with GPS spoofing attacks that has been shown in an experiment using LabSat3 GPS simulator [10,57].

An attacker can make use of Nmap that stands for network mapper. The attacker is capable of identifying, the open port and is currently in running state to exploit the loophole and compromise the entire control system. Nmap is useful to snoop into the wireless network and allow capturing packets just like the tool, Wireshark. For instance, open file transfer protocol FTP port allows to connect with Wi-Fi and login to FTP to steal very personal information regarding flight data, media, etc., giving beneficial insights to file system [55,58]. Also, it is explored in aforementioned section that how spoofing can be done using various means that shows the importance of applying more robust and effective methodologies to safeguard unmanned aerial CPS.

7.3.5.1 Rogue Updates

A network attack where an attacker deliberately introduces rogue device or rogue nodes to the interconnected network, posing as a legitimate node, leverages its capabilities to read all communications in the network, push messages and illegitimate files to other nodes in the network. Sometimes software updates for control system in a UAV need to be done and this opportunity is taken by the attackers to introduce malicious and illegitimate updates in the software which can further result in the malfunctioning of entire UAV. A credible solution is FOTA (Firmware updates Over the Air) [25].

7.3.5.2 OBD Port Exploitation

On-board Diagnostics (OBD) is physically located in a car that controls ECU and OBD and has direct access to network infrastructure. If OBD is exploited, then the attacker can have direct access to the entire network infrastructure. OBD can modify code-base which is responsible for engine, lighting and breaking functionality. Also, OBD is responsible for maintenance of ECU and upgrading ECU [25].

7.3.5.3 Close Proximity Vulnerabilities

This kind of vulnerability can be exploited when exposed via short-range communication and exploitation of on-board sensors. More planned and sinister attacks include a malicious user who keeps on sending false signals to the vehicle with an intention to distort, hijack and restrict it with a virtual fence. Signal jamming works by deliberately blocking, jamming or interfering with the authorized communication among connected devices [50]. It also decreases the signal to noise ratio. Inter-vehicle communication can be broken or interrupted through signal jamming that can further result in DoS attack. Jamming can be of three types – trivial jamming in which attacker constantly transmits the noise, periodic attack that involves noise broadcast in random directions and at random intervals, and reactive attack where noise broadcast occurs when communication signals are detected [59].

Possible solution of jamming attack can assimilate the detection of intended signal in environments where there is a large volume of noise, techniques adopted from the UAVs including radio communication are of utmost importance for safe utilization and functioning of UAVs and to identify the signal amongst large volumes of noise that can include different identification, filtering and signal processing techniques. To prevent spoofing attacks and GPS jamming attack, integrated sensors can be used. Also, multiple systems when fused together, can improve reliability [49, 60].

Hacking and infiltrating CPS, including its design, main function and components affects the responsibility, liability, data ownership and privacy of systems. Sensors and actuators handling task including network embedded systems are connected with each other through internet. When autopilot mode is switched on then engine, position and speed need to be controlled. Then everything is done by actuators that trigger to address mental adjustments to meet the settings of autopilot [61].

7.4 PENETRATION TESTING WITH KALI LINUX

Three aspects have been considered in order to proceed with proposed methodology, including detection, location and mitigation of DoS attack.

For demonstrating the possible ways of launching an attack on an unmanned aerial-ship, we have used Kali Linux, that is a Linux-based OS that contains a suite for penetration tools. We have empaneled below some penetration tools that have been acclimatized according to our requirements, since we didn't have funding for this project and we had to make use of available resources. After performing a detailed survey on different tools available to perform penetration testing, we inferred that Kali Linux along with its other tools can be used as a substitute for demonstrating attacks on UAVs. As the rudimentary parameters remain the same such as IPs, data transmission model, protocols such as TCP/UDP, DHCP, ARP, etc. are used [62].

We used the latest version of Kali Linux, that is, 2020.2. We installed it on a Virtual Machine box, in a MacOS environment.

7.4.1 DETECTION OF DOS ATTACK

DoS attacks mostly rely on rudimentary vulnerabilities in a computational infrastructure or a network that can include unpatched systems, absence of authentication, existence of reflectors or amplifiers, poorly configured systems, including virtual cyber space and lack of knowledge to identify an attack [50]. A qualitative view of information risk (also a measure of cyber-attack lethality) in a system such as SAA or computer network is expressed as [45]:

$$\text{Risk} = \big(\text{Threats} \times \text{Vulnerabilities} \times \text{Impact/Countermeasures}\big) \quad (7.1)$$

And at time state 0, this equation can be reduced to:

$$\text{Risk} = \text{function}\big(\text{Threats / Countermeasures}\big) \quad (7.2)$$

At time state $=0$, where vulnerabilities & impact are constants and drop out of the equation.

DNS Spoofing attack is also called DNS cache poisoning. It is a kind of computer hacking where a corrupt Domain Name System data can be introduced into the DNS resolver's cache, causing the name server to return an incorrect IP address that can result in traffic being diverted to the attacker's machine. The main variant of DNS cache poisoning includes redirection of name server of adversary's domain to the name server of victim's domain, then assigning that name server an IP address specified by the attacker. No firewall or antivirus can detect such attacks easily. Since a ground control station remains in a continuous contact with remotely controlled aircraft to monitor the position, velocity and the particular time at which the UAV is flying, therefore these three parameters can be used to detect the launched attack (Figure 7.7).

Time: Whenever a UAV is in motion, it gets a request from the control station to change its altitude, direction or position in a specific time frame to monitor the navigation of UAV. However, when it comes to a DoS attack, this motion request can come in short intervals of time over and over that can perplex its navigational senses. Assessing the overall counts of requests coming to the navigational sensor of a UAV, a DoS attack can be detected with much ease [63].

Velocity: Since the users monitoring a UAV from a ground control station know the exact speed at which the unmanned aircraft is flying such as 3 km/s or 2,000 km/h, it gets convenient for them to determine their exact arrival time at

FIGURE 7.7   Network flooding.

the destination. Now, if a UAV takes a detour or does not follow the original path, then its velocity can show fluctuations along with its arrival time. The velocity or angular velocity of a UAV is always synchronized with time and displacement, and if the unmanned drone is under a cyber-attack which is affecting the velocity of UAV showing frequent changes in the direction of drone on remote controller's screen, then it can be used to identify that a spoofed attack has been launched on a particular UAV [64].

Position: An unmanned airplane that is being monitored by a ground control station will definitely be affected by its sudden change in displacement that can affect the coordinates of the unmanned drone. The drone which is under a constant surveillance of control station will know immediately that an unplanned displacement with directional changes of the UAV is nothing but a spoofed cyber-space attack.

7.5 LOCATION OF DoS ATTACK

To detect a GPS spoofing attack, the incoming packet needs to be inspected so that useful details like source address, destination address, sequence number, hop counts and payload that can be utilized to predict the provenance of attack so as to track down the attacker. Nowadays, LASER is used to provide elevation information with higher accuracy to improve GNSS navigational system. Besides, Navy uses laser gyroscopes, where there is no GPS signal availability [65].

The spoofed IP address can be tracked down, using a networked GPS and GNSS that can be used further to get the measurements of the position of the attacker's unmanned drone. These laser gyrocompasses can be upgraded and brought into use for detecting the location of other UAV that is launching the attack. Also, Laser Range Finders (LRF) can be used to determine accurate range and angle information [62].

In order to use a laser in a combination with the networked GPS, we need to perform some mathematical formulations and development of executable algorithms is a must that can trace the IP address and location of the attacker's UAV both in minimum time with no time delays at all since these armed UAVs have flight time constraints.

7.6 DEMONSTRATION OF DoS ATTACKS ON UAVS

Here, we are considering a DoS attack that can be launched from any part of this planet, using a wireless connection of cyber-space world. There are various parameters of a UAV that can be utilized to predict that an attack has been launched on an unmanned aircraft.

7.6.1 MITM ATTACK USING ETTERCAP ON KALI LINUX BY ARP POISONING

7.6.1.1 Locating Open Ports

Below are some Nmap commands that can be used on an identified host in a specified network that can assist in locating open ports which can be exploited by the attacker.

Also, if there is any firewall or IDS embedded in system to identify any incoming threat, it can be evaded and spoofed using Nmap.

7.6.1.2 NMAP

TARGET SPECIFICATION:

Can pass hostnames, IP addresses, networks, etc.
Ex: scanme.nmap.org, microsoft.com/24, 192.168.0.1; 10.0.0–255.1–254

Ettercap is a comprehensive suite to perform MITM attacks on local network. It sniffs live connections, content filtering and much more. It supports both active and passive dissection of various protocols. Ettercap 0.8.3(EB) is used to perform this attack.

- Target/victim machine = Apple MacOS
- Attacking machine = Kali Linux OS

Targets 1 and 2, where target 1 will be the router's IP address and target 2 will be the Victim's IP address. Figure 7.8 depicts port stealing on victims' network and targets that have been ARP poisoned. Since ARP poisoning allows you to spoof ARP of a device, hence all the incoming traffic on target's network will be redirected to adversary's network that will become a vantage point for the attacker and he can harm the victim in multifarious ways [66].

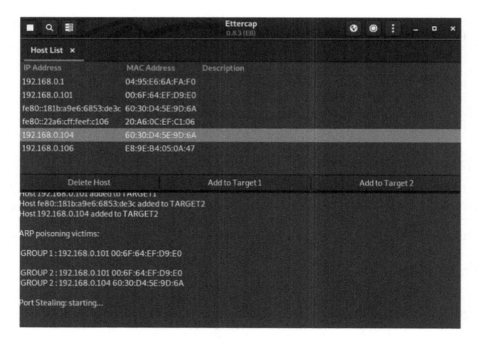

FIGURE 7.8   MITM attack using Ettercap.

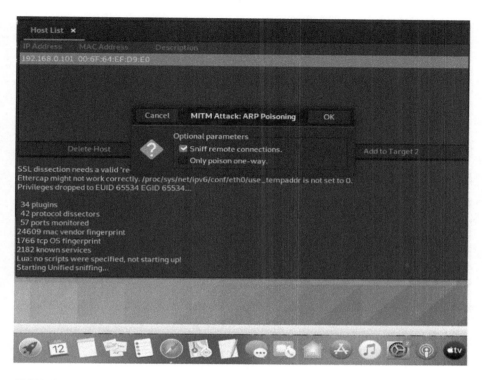

FIGURE 7.9 Unified sniffing on remote connections using MITM attack, using ARP spoofing technique.

Figure 7.9 shows unified sniffing on remote connections using MITM attack, using ARP spoofing technique. It also represents that currently during the attack, nearly 57 ports are being monitored. Since we have used MacOS as the victim OS, there are 24609 mac vendor fingerprints under surveillance.

More options that can be explored on Kali are mentioned below [66]:

FIREWALL/IDS EVASION AND SPOOFING:
  -f; --mtu <val>: fragment packets (optionally w/given MTU)
  -D <decoy1,decoy2[,ME],...>: Cloak a scan with decoys
  -S <IP_Address>: Spoof source address
  -e <iface>: Use specified interface
  -g/--source-port <portnum>: Use given port number
  --proxies <url1,[url2],...>: Relay connections through HTTP/SOCKS4
      proxies
  --data <hex string>: Append a custom payload to sent packets
  --data-string <string>: Append a custom ASCII string to sent packets
  --data-length <num>: Append random data to sent packets

Figure 7.10 demonstrates how packets can be captured using Wireshark and Ettercap. Wireshark is used to capture packets of any kind on any port and Ettercap can be

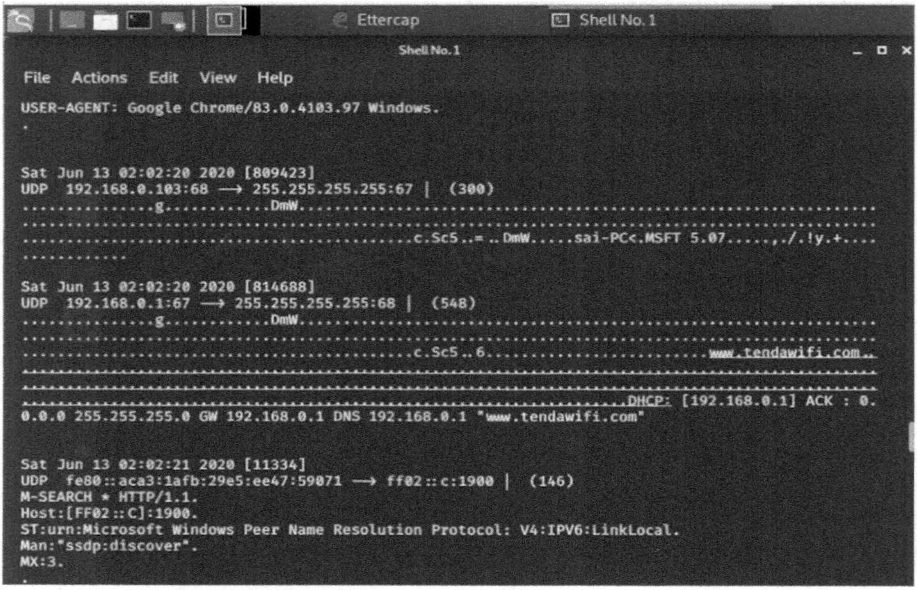

FIGURE 7.10   Packets captured using Wireshark and Ettercap

used to spoof the ARP or IP address of that specific packet. Below, you can find some scanning techniques including cookie-echo scan, TCP scan, idle scan and many more can be explored using Kali [51]:

SCAN TECHNIQUES:
-sS/sT/sA/sW/sM: TCP SYN/Connect ()/ACK/Window/Maimon scans
-sU: UDP Scan
-sN/sF/sX: TCP Null, FIN, and Xmas scans
--scanflags <flags>: Customize TCP scan flags
-sI <zombie host[:probeport]>: Idle scan
-sY/sZ: SCTP INIT/COOKIE-ECHO scans

Hence, it is crystal clear that combining various tools in kali forms a lethal combination and an adversary can make attempts on attacking a UAV or a drone. We need to analyze the existing approaches in detail and require more scholars and researchers who can contribute in enhancing the security of these unmanned battleships. Based on the aforementioned experiments with Kali Linux pen-testing tool, it can be said that we can attack on drones by finding information about their network infrastructure. This empirical analysis demands us to concentrate more on reinforcement of cyber security tactics that includes using strong underlying infrastructure, foolproof protocols, updated software, VPNs, DMZs, strong firewalls, multi-level authentication and much more.

7.7 CONCLUSION AND FUTURE SCOPE

To recapitulate, after going through various research papers, including different methodologies, we were able to come up with some of the techniques to cope up with the most disrupting attacks i.e., DoS attack, where the legitimate user has no control over the unmanned drone leading to disastrous and dire consequences. With the existence of multifarious techniques and methodologies to cope up with DoS attack, it gets more befuddling that which technique should be adopted and how to proceed with a particular methodology that may or may not be compatible with the aspects of dealing with detection and location of a DoS attack. For detection and location of an attacker in order to identify the spoofed IP-based DoS attack, the development of a much efficient and effective algorithm designed in collaboration with laser and networked GPS is much needed.

For future directions, aspirants can work on mitigation of DoS attacks with the concept of advanced whitelisting and blacklisting concepts, and hashing function based advanced hash tables. Moreover, the propounded research is also flexible enough to explore several methods, for instance, the model can be enhanced to be employed for prevention against Distributed Denial of Service (DDoS) attacks by considering a greater number of sources of attacks generated that include ample number of spoofed IP packets. Keeping the baseline parameters, several techniques can be explored, including fuzzy logics, packet filtering techniques, kernel inspection methods, IDS with entropy-based systems and much more to cope with DDoS attacks on a UAV.

REFERENCES

1. Grieves, M. (2014). Digital twin: Manufacturing excellence through virtual factory replication, *A White Paper*; Michael Grieves, LLC: Melbourne, FL, USA.
2. Rasheed, A., San, O., & Kvamsdal, T. (2019). Digital twin: Values, challenges and enablers. arXiv preprint arXiv: 1910.01719.
3. Madni, A. M. (2017). *Transdisciplinary Systems Engineering: Exploiting Convergence in a Hyper-Connected World*; Foreword by Norm Augustine. Springer, Berlin, Germany.
4. Glaessgen, E. H., & Stargel, D. S. (2012, April). The digital twin paradigm for future NASA and U.S. Air Force Vehicles. In *53rd AIAA/ASME/ASCE/AHS/ASC Structures, Structural Dynamics, and Materials Conference 20th AIAA/ASME/AHS Adaptive Structures Conference 14th AIAA*, Honolulu, HI, USA, 23–26 April 2012 (p. 1818).
5. Hearn, M., & Rix, S. (2019). Cybersecurity considerations for digital twin implementations. *IIC Journal of Innovation*. 107–113.
6. Yang, W., Tan, Y., Yoshida, K., & Takakuwa, S. (2017). *Digital Twin-Driven Simulation for a Cyber-Physical System in Industry 4.0*. DAAAM International Scientific Book (pp. 227–234). Vienna, Austria
7. Sabatini, R., Gardi, A., & Richardson, M. (2014). LIDAR obstacle warning and avoidance system for unmanned aircraft. *International Journal of Mechanical, Aerospace, Industrial and Mechatronics Engineering*, 8(4), 718–729.
8. Sabatini, R., Moore, T., & Hill, C. (2014). GNSS avionics-based integrity augmentation for RPAS detect-and-avoid applications. In: Fourth Australasian Unmanned Systems Conference, 2014 (ACUS 2014), 10-12 Dec 2014, Melbourne, Australia.
9. Watkins, L., Ramos, J., Snow, G., Vallejo, J., Robinson, W. H., Rubin, A. D., ... Li, C. (2018, June). Exploiting multi-vendor vulnerabilities as back-doors to counter the threat of rogue small unmanned aerial systems. In *Proceedings of the 1st ACM MobiHoc Workshop on Mobile IoT Sensing, Security, and Privacy* (p. 1). ACM.

10. Dey, V., Pudi, V., Chattopadhyay, A., & Elovici, Y. (2018, January). Security vulnerabilities of unmanned aerial vehicles and countermeasures: An experimental study. In *VLSI Design and 2018 17th International Conference on Embedded Systems (VLSID), 2018 31st International Conference on* (pp. 398–403). IEEE.

11. Fanelli, R., & Conti, G. (2012, June). A methodology for cyber operations targeting and control of collateral damage in the context of lawful armed conflict. In *Cyber Conflict (CYCON), 2012 4th International Conference on* (pp. 1–13). IEEE.

12. Mohurle, S., & Patil, M. (2017). A brief study of wannacry threat: Ransomware attack 2017. *International Journal of Advanced Research in Computer Science*, 8(5). 1938–1940.

13. Guvenc, I., Koohifar, F., Singh, S., Sichitiu, M. L., & Matolak, D. (2018). Detection, tracking, and interdiction for amateur drones. *IEEE Communications Magazine*, 56(4), 75–81.

14. Madni, A. M., Madni, C. C., & Lucero, S. D. (2019). Leveraging digital twin technology in model-based systems engineering. *Systems*, 7(1), 7.

15. Fakeeh, K. A., & King, J. (2016). An overview of DDOS attacks detection and prevention in the cloud. *Network*, 11(7). International Journal of Applied Information Systems (IJAIS) – ISSN : 2249-0868 Foundation of Computer Science FCS, New York, USA, 25–34

16. Wang, E. K., Ye, Y., Xu, X., Yiu, S. M., Hui, L. C. K., & Chow, K. P. (2010, December). Security issues and challenges for cyber physical system. In *Green Computing and Communications (GreenCom), 2010 IEEE/ACM Int'l Conference on & Int'l Conference on Cyber, Physical and Social Computing (CPSCom)* (pp. 733–738). IEEE.

17. Sabatini, R., Gardi, A., & Ramasamy, S. (2014). A laser obstacle warning and avoidance system for unmanned aircraft sense-and-avoid. *Applied Mechanics and Materials*, 629, 355–360.

18. Rudinskas, D., Goraj, Z., & Stankūnas, J. (2009). Security analysis of uav radio communication system. *Aviation*, 13(4), 116–121.

19. Calnoor Rajashekar, S., Tourani, R., & Gururajan, S. (2020). Secure communications in unmanned aerial system swarms. In *AIAA Aviation 2020 Forum* (p. 2926).

20. Khan, N. A., Brohi, S. N., & Jhanjhi, N. Z. (2020). UAV's applications, architecture, security issues and attack scenarios: A survey. In *Intelligent Computing and Innovation on Data Science* (pp. 753–760). Springer, Singapore.

21. Wen, G., Yu, W., Yu, X., & Lü, J. (2017). Complex cyber-physical networks: From cybersecurity to security control. *Journal of Systems Science and Complexity*, 30(1), 46–67.

22. Meredith, J., Straub, J., & Bernard, B. (2019, December). Identifying UAV swarm command methods and individual craft roles using only passive sensing. In *2019 International Conference on Computational Science and Computational Intelligence (CSCI)* (pp. 1043–1046). IEEE.

23. Siddappaji, B., & Akhilesh, K. B. (2020). Role of cyber security in drone technology. In *Smart Technologies* (pp. 169–178). Springer, Singapore.

24. Khan, S., Farnsworth, M., McWilliam, R., & Erkoyuncu, J. (2020). On the requirements of digital twin-driven autonomous maintenance. *Annual Reviews in Control*, 50, 13–28.

25. Parkinson, S., Ward, P., Wilson, K., & Miller, J. (2017). Cyber threats facing autonomous and connected vehicles: Future challenges. *IEEE Transactions on Intelligent Transportation Systems*, 18(11), 2898–2915.

26. Haque, M. S., & Chowdhury, M. U. (2019, November). Ad-Hoc framework for efficient network security for Unmanned Aerial Vehicles (UAV). In *International Conference on Future Network Systems and Security* (pp. 23–36). Springer, Cham.

27. Shivers, M., Llanes, C., & Sherman, M. (2019, November). Implementation of an artificial immune system to mitigate cybersecurity threats in unmanned aerial systems. In *2019 IEEE International Conference on Industrial Internet (ICII)* (pp. 12–17). IEEE.

28. Wei, Y., Hong, T., & Kadoch, M. (2020). Improved Kalman filter variants for UAV tracking with radar motion models. *Electronics*, 9(5), 768.

29. Zhou, L., Liao, M., Yuan, C., & Zhang, H. (2017). Low-rate DoS attack detection using expectation of packet size. *Security and Communication Networks, 2017.*

30. Liang, C., Wen, F., & Wang, Z. (2019). Trust-based distributed Kalman filtering for target tracking under malicious cyber attacks. *Information Fusion*, 46, 44–50.

31. He, D., Chan, S., Qiao, Y., & Guizani, N. (2018). Imminent communication security for smart communities. *IEEE Communications Magazine*, 56(1), 99–103.

32. Ramasamy, S., & Sabatini, R. (2015, June). A unified approach to cooperative and non-cooperative sense-and- avoid. In *Unmanned Aircraft Systems (ICUAS), 2015 International Conference on* (pp. 765–773). IEEE.

33. Monostori, L. (2014). Cyber-physical production systems: Roots, expectations and R&D challenges. *Procedia CIRP*, 17, 9–13.

34. Harvey, A. C. (1990). *Forecasting, Structural Time Series Models and the Kalman Filter.* Cambridge University Press. UK

35. Zekavat, R., & Buehrer, R. M. (2011). *Handbook of Position Location: Theory, Practice and Advances* (Vol. 27). John Wiley & Sons. Hoboken, New Jersey

36. Cardenas, A., Amin, S., Sinopoli, B., Giani, A., Perrig, A., & Sastry, S. (2009, July). Challenges for securing cyber physical systems. In *Workshop on Future Directions in Cyber-Physical Systems Security* (Vol. 5).

37. He, D., Qiao, Y., Chan, S., & Guizani, N. (2018). Flight security and safety of drones in airborne fog computing systems. *IEEE Communications Magazine*, 56(5), 66–71.

38. Madni, A. M. (2015). Expanding stakeholder participation in upfront system engineering through storytelling in virtual worlds. *Systems Engineering*, 18, 16–27.

39. Madni, A. M., Spraragen, M., & Madni, C. C. (2014). Exploring and assessing complex system behavior through model-driven storytelling. In *Proceedings of the IEEE Systems, Man and Cybernetics International Conference, Invited Special Session "Frontiers of Model Based Systems Engineering"*, San Diego, CA, USA, 5–8 October 2014.

40. Sun, X., Ng, D. W. K., Ding, Z., Xu, Y., & Zhong, Z. (2019). Physical layer security in UAV systems: Challenges and opportunities. *IEEE Wireless Communications*, 26(5), 40–47.

41. Sabatini, R., Richardson, M. A., Gardi, A., & Ramasamy, S. (2015). Airborne laser sensors and integrated systems. *Progress in Aerospace Sciences*, 79, 15–63.

42. Director, J. S. (2013). *Joint Publication 3–12 (R) Cyberspace Operations.* Joint Chiefs of Staff, Washington.

43. Chao, H., Cao, Y., & Chen, Y. (2010). Autopilots for small unmanned aerial vehicles: A survey. *International Journal of Control, Automation and Systems*, 8(1), 36–44.

44. Tanil, C., Khanafseh, S., & Pervan, B. (2015, September). GNSS spoofing attack detection using aircraft autopilot response to deceptive trajectory. In Proceedings of the 28th International Technical Meeting of the Satellite Division of the Institute of Navigation (*ION GNSS+2015*), Tampa, FL, (pp. 3345–3357).

45. Nichols, R. K., Ryan, J. J. C. H., Mumm, H. C., Lonstein, W. D., Carter, C., & Hood, J. P. (2019). Understanding hostile use and cyber-vulnerabilities of UAS: Components, autonomy v automation, sensors, SAA, SCADA and cyber attack taxonomy. *Unmanned Aircraft Systems in the Cyber Domain.* https://kstatelibraries.pressbooks. pub/unmannedaircraftsystems/chapter/chapter-3-understanding-hostile-use-and-cyber-vulnerabilities-of-uas-components-autonomy-v-automation-sensors-saa-scada-and-cyber-attack-taxonomy/

46. Azzabi, T., Farhat, H., & Sahli, N. (2017, January). A survey on wireless sensor networks security issues and military specificities. In *Advanced Systems and Electric Technologies (IC_ASET), 2017 International Conference on* (pp. 66–72). IEEE.

47. Mairaj, A., Majumder, S., & Javaid, A. Y. (2019, June). Game theoretic strategies for an unmanned aerial vehicle network host under DDoS attack. In *2019 International Conference on Unmanned Aircraft Systems (ICUAS)* (pp. 120–128). IEEE.

48. Sanjab, A., Saad, W., & Başar, T. (2019). A game of drones: Cyber-physical security of time-critical UAV applications with cumulative prospect theory perceptions and valuations. arXiv preprint arXiv:1902.03506.

49. Manesh, M. R., Kenney, J., Hu, W. C., Devabhaktuni, V. K., & Kaabouch, N. (2019, January). Detection of GPS spoofing attacks on unmanned aerial systems. In *2019 16th IEEE Annual Consumer Communications & Networking Conference (CCNC)* (pp. 1–6). IEEE.

50. Zhou, L., Liao, M., Yuan, C., & Zhang, H. (2017). Low-rate DoS attack detection using expectation of packet size. *Security and Communication Networks*, 2017.

51. Tegeler, F., Fu, X., Vigna, G., & Kruegel, C. (2012, December). Botfinder: Finding bots in network traffic without deep packet inspection. In *Proceedings of the 8th International Conference on Emerging Networking Experiments and Technologies* (pp. 349–360).

52. Span, M., Mailloux, L. O., Mills, R. F., & Young, W. (2018). Conceptual systems security requirements analysis: Aerial refueling case study. *IEEE Access*, 6, 46668–46682.

53. Donkal, G., & Verma, G. K. (2018). A multimodal fusion based framework to reinforce IDS for securing Big Data environment using Spark. *Journal of Information Security and Applications*, 43, 1–11.

54. Mukherjee, S., Shirazi, H., Ray, I., Daily, J., & Gamble, R. (2016, December). Practical DoS attacks on embedded networks in commercial vehicles. In *International Conference on Information Systems Security* (pp. 23–42). Springer, Cham.

55. Javaid, A. Y., Jahan, F., & Sun, W. (2017). Analysis of Global Positioning System-based attacks and a novel Global Positioning System spoofing detection/mitigation algorithm for unmanned aerial vehicle simulation. *Simulation*, 93(5), 427–441.

56. Dahiya, S., & Garg, M. (2019, April). Unmanned aerial vehicles: Vulnerability to cyber attacks. In *International Conference on Unmanned Aerial System in Geomatics* (pp. 201–211). Springer, Cham.

57. Nunez, J., Tran, V., & Katangur, A. (2019). Protecting the unmanned aerial vehicle from cyberattacks. In *Proceedings of the International Conference on Security and Management (SAM)* (pp. 154–157). The Steering Committee of The World Congress in Computer Science, Computer Engineering and Applied Computing (WorldComp).

58. He, D., Chan, S., & Guizani, M. (2017). Drone-assisted public safety networks: The security aspect. *IEEE Communications Magazine*, 55(8), 218–223.

59. Guo, R., Wang, B., & Weng, J. (2020, April). Vulnerabilities and attacks of UAV cyber physical systems. In *Proceedings of the 2020 International Conference on Computing, Networks and Internet of Things* (pp. 8–12).

60. Sathyamoorthy, D. (2015). A review of security threats of unmanned aerial vehicles and mitigation steps. *The Journal of Defense and Security* (In press), 6(2).

61. Nguyen, P. H., Ali, S., & Yue, T. (2017). Model-based security engineering for cyber-physical systems: A systematic mapping study. *Information and Software Technology*, 83, 116–135.

62. Owens, K. The Navy is using laser gyroscope tech to navigate in GPS-denied environments. https://defensesystems.com/articles/2017/09/19/northrop-navy-gyro.aspx. Accessed on 21st December 2019.

63. Vasconcelos, G., Miani, R. S., Guizilini, V. C., & Souza, J. R. (2019). Evaluation of DoS attacks on commercial wi-fi-based UAVs. *International Journal of Communication Networks and Information Security*, 11(1), 212–223.

64. Park, S., Jung, J., Oh, S., Lee, W., & Kim, H. (2019, January). Integrated cyber-physical attack detection and response system for resilient multi-UAV control. In *7th Asian/Australian Rotorcraft Forum*, ARF 2018.
65. Bremler-Barr, A., Harchol, Y., Hay, D., & Koral, Y. (2014, December). Deep packet inspection as a service. In *Proceedings of the 10th ACM International on Conference on Emerging Networking Experiments and Technologies* (pp. 271–282). ACM.
66. https://www.kali.org. Accessed on 1st April 2020.

8 Digital Twin Techniques in Recognition of Human Action Using the Fusion of Convolutional Neural Network

Meenu Gupta, Rakesh Kumar, and Sapna Dewari
Chandigarh University

CONTENTS

DOI: 10.1201/9781003132868-8

8.1   INTRODUCTION

The digital twin has increased the curiosity of many researchers recently. It produces a copy of the physical world into the virtual world that works the same way as the original product performs in real-time. If there is an adjustment in the virtual room, the same adjustment will implicate the real project in real-time or vice-versa. The digital twin technology improves ongoing operation, tests the new product, increases efficiency, and saves time and money. Digital twin technology has much application in the real world as in industry, healthcare, agriculture, aerospace, etc. Here, recognition of action plays a crucial role in receiving changes in the physical space, to know the actions a human performs.

In a real-world setting, action recognition has been an active field of research for decades as shown in Table 8.1 and Figure 8.1. It addresses many subjects, such as video human identification, human pose estimation, human tracking, and time series interpretation and cognition. The current study focuses on realistic datasets gathered from movies, web videos, television programs, etc. Action identification has various benefits, such as monitoring human emotions, detecting appropriate activity, and many more. But in the domain of computer vision and machine learning, obtaining

TABLE 8.1

Tabular Representation of Human Action Recognition on a Different Platform

Research Journals	Results
IEEE	4,727
ELSEVIER	11,815
Springer	47,942
Hindawi	10,000
ScienceDirect	**29,6163**
Microsoft Academic	2,106
Core	16,736,177

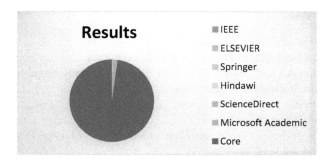

FIGURE 8.1   Result for human action recognition on a different platform.

appropriate facts is difficult. For the best result, one must know about the behavior and effects of the human surrounding [1].

To understand the conduct, one must feed relevant data into the machine that improves with the preface of RGB cameras and depth cameras. Although connectivity identification and action classification are two primary components within the actual scenario, the connectivity applies to act, combining various persons or events, including holding a gun or playing an instrument, between entities and objects. Identification of action refers to locating the point at which an event occurred from the picture ordered items in time and space.

The resulting systems show competitive performance in visual object recognition, human action recognition, brain-computer interconnection, human tracking, etc. Different approaches (i.e., ANN, CNN, RNN, etc.) to extract and classify these features for recognizing human action have been conducted [2]. Figure 8.1 shows the trend of human action recognition in several research platforms. Here, we will focus on human action extraction and classify them using different deep learning approaches like CNN, RNN, ANN, etc.

We present digital twins and the awareness of human behavior in Section 8.1 and a brief description about the chapter. Section 8.2 gives us an insight into the route of action recognition. Afterward, we will discuss the relevant works in understanding human behavior and the digital twin in Section 8.3. We will address the architecture of recognizing human behavior with a digital twin layer in Section 8.4. Then, we will discuss multiple neural networks used in human action extraction and classification in Section 8.5. Section 8.6 will include case studies involving identification of action and the digital twin. We'll eventually address the conclusion of this document.

8.2 HISTORY OF HUMAN ACTION RECOGNITION AND DIGITAL TWIN

Early studies were motivated by human representation in the arts by Da Vinci. According to him, human actions are characterized by how different body parts move relative to every other. Familiarity with the anatomy of nerves, bones, muscles, and sinews helps to understand various motions with their levels of strength [3]. Giovanni Alfonso Borelli wrote about spinal biomechanics that contributed to ascertain in-depth and undiminished accuracy [4]. Etienne-Jules Marey contributed significantly in the emerging field of cinematography [5]. Eadweard Muybridge also contributed by inventing the machine for displaying the recorded series of images. He applied his techniques for movement studies [6]. Gunnar Johanson used image order for programmed human motion analysis [7]. Johansson showed that human observers can recognize biological motion patterns with moving light displays (MLD), attached to the body parts, even when presented with few moving dots [8]. Loffe and Forsyth represent people as collections of nine body segments, one for the torso and two for each limb to locate people [9]. Later, image segmentation based on skin-color and shape analysis and the invariant moments are combined. The features are extracted and used for an input vector to a radial basis function network (RBFN) [10]. Real-time analysis of video stream is used to detect human motion and extract its boundaries [2]. A human activity analysis was done in a complex environment [11]. The actions of humans from still images were recognized [12]. Recognizing humans' action in 3-D through skeletons was proposed [13]. A multimodal dataset was used for human action recognition that utilized a depth camera and a wearable inertial sensor [14] along with representation learning of temporal dynamics for skeleton-based action recognition [15]. Task-based control and human activity recognition for human-robot collaboration [16] was conducted. Deep CNN-based data-driven recognition of cricket batting shots [17] was also conducted. WiAct is a passive wi-fi based human activity recognition system that works on various neural networks [18,19]. A CSI-based system using Wi-fi for human activity recognition was conducted to analyze common human behaviors [20]. A novel graph edge convolution was used to represent an edge by combining every neighboring border [21]. A combined strategy wherein many trained networks are dependent on the learning method, recognizing human activity with the low-power multipath deep neural network on event camera outputs [22], motion edge highlighted optical flow method [23], etc. were utilized. Still, many other types of researches are going on to increase the performance of human action recognition. Figure 8.2 shows the changes shown in the field of recognizing human action. Recently, the act of recognizing humans is widely used in digital twin environment. Human updation made in the physical world needs updation in twin environments for better results in the area like production, disaster management, and many more [24–26].

8.3   RELATED WORK

The human action recognition technique flourishes in fields like video surveillance system, robotics for human behavior characteristics, multiple activity recognition systems, and recently in the digital twin environment. Different researchers have studied various parts to improve human action recognition and have made remarkable progress as mentioned below.

Shotton et al. [27], proposed a method for quickly and accurately estimating 3D positions of skeletal joints using a single depth image. Here, skeleton-based action recognition approaches can be grouped into two main categories: joint-based and body part-based.

Vemulapalli et al. [13], inspired by relative geometry between various body components, provided a more meaningful description than their absolute locations. They used the MSR-Action dataset and UTkinect-Action dataset for their experiment.

Du et al. [28] proposed an end-to-end hierarchical recurrent neural network for skeleton-based action recognition. Five different deep RNN architectures were derived from their proposed model and were compared to verify the proposed network. Datasets used were MSR Action3D, Motion Capture, HDM05, and Berkeley MHAD. Yong Du, et al. [15] proposed an end-to-end hierarchical recurrent neural network for extracting temporal dynamic of skeleton orders and to recognize the corresponding actions. To improve viewpoint variation, random scale and rotation transformations were adopted during training.

In Ref. [29], Mahshid Majd and Reza Safabakhsh proposed to change LSTM units to cover all three types of information needed to categorize action classes in a video. Convolution and correlation operators are used to form a part, capable of

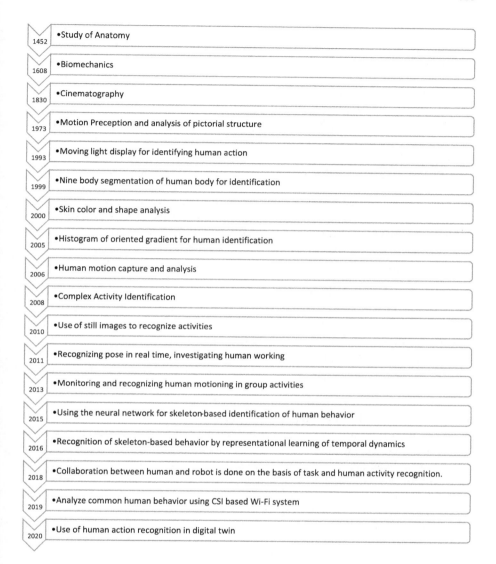

1452	•Study of Anatomy
1608	•Biomechanics
1830	•Cinematography
1973	•Motion Preception and analysis of pictorial structure
1993	•Moving light display for identifying human action
1999	•Nine body segmentation of human body for identification
2000	•Skin color and shape analysis
2005	•Histogram of oriented gradient for human identification
2006	•Human motion capture and analysis
2008	•Complex Activity Identification
2010	•Use of still images to recognize activities
2011	•Recognizing pose in real time, investigating human working
2013	•Monitoring and recognizing human motioning in group activities
2015	•Using the neural network for skeleton-based identification of human behavior
2016	•Recognition of skeleton-based behavior by representational learning of temporal dynamics
2018	•Collaboration between human and robot is done on the basis of task and human activity recognition.
2019	•Analyze common human behavior using CSI based Wi-Fi system
2020	•Use of human action recognition in digital twin

FIGURE 8.2   Evolution of human action recognition.

extracting spatial and motion features and their temporal dependencies. They tested the proposed unit called CCLSTM, in a deep architecture with convolutional layers on UCF101 and HMDB51 datasets.

Sahoo et al. [30] proposed a technique to extract features from action videos by HOG and also proposed to use BoHOF with the SVM classifier for the KTH dataset. In Ref. [31], NTU RGB+D is a state-of-the-art, large-scale benchmark for action recognition. It illustrates a series of standards and experience for large-scale data building. KETI RGB+D is a dataset optimized for video surveillance applications.

We trained the proposed CNN with the NTU RGB+D dataset and KETI dataset. KETI dataset used for training has 20 video clips, where each video chip consists of four-predefined front view action images. It used Kinect video camera as an input device and recognition results as output. The system can process 20 frames per second of video input. The proposed method performs in-depth gradient-based feature construction and CNN-based action recognition.

In Ref. [32], they built a CNN model to predict the confidence maps for identifying the body part. First, 2D skeleton points were recognized and utilized to form complex architecture. Then, time convolution at 2D skeleton points was performed to accurately predict 3D posture in the video. NTU-RGB+D 3D skeleton dataset was chosen for testing the proposed work. The ST-GCN was proposed, which achieved remarkable results. However, there are still challenges in practical application scenarios.

Except for joints, Xikun Zhang et al. [19] noted that the movement of limbs is important to understand behavior. Given this finding, for skeleton-based action identification, they explored the dynamics of human limbs. In a graph for the human skeleton, they represented an edge by combining its spatial coordinates to encode the coordination between different wings and temporal-adjacent margin to achieve stable movements of motion. By adding various mutual intermediate layers to combine graph node and edge CNNs, they further created two-hybrid networks. Graph edge convolution and hybrid network integration have been tested on Kinetics and NTURGB+D data sets for edge convolution and traditional node convolution. To prove the principle of adaptive route planning through a dynamic protective cover using a digital twin approach, Klaus Dröder et al. [33] suggested an experimental simulation framework for human-robot interaction.

In Ref. [34], they recommended a p-non local block to solve the far region under the hypothesis of deducting computational complexity. The Bi-LSTM and KeyLess Attention are arranged to generate RoI of human motion. A specific approach for human behavior detection and recognition is illustrated by Nadeem et al. [35], with linear discriminant analysis and artificial neural networks. Multifaceted characteristics with linear discriminant analysis have been created by human silhouette and body parts that detect human behavior. The Weizmann Human Action dataset and KTH dataset were used to verify the accuracy of the work proposed.

An assembly-commissioning complete factor knowledge model based on digital twin technology was constructed by Sun Xuemin et al. [36]. They merged twin data and the relationship between the asset-liability prediction and assembly-commissioning quality management digital twin model.

To summarize and display the outcomes, we have presented them in a more crystalized way in Table 8.2.

8.4 ARCHITECTURAL FRAMEWORK OF HUMAN ACTION RECOGNITION IN DIGITAL TWIN TECHNOLOGY

The digital twin technology allows the physical and virtual world to communicate, as shown in Figure 8.3. The concept behind it is that it provides a platform where the non-living objects can show a living behavior. For example, if we build a twin

TABLE 8.2
Experimental Result Acquired Using Different Models

Paper	Highlight	Method	Dataset	Outcomes and Future Work
Shotton et al. (2013) [27]	From single depth image prediction of human pose, it is done with the 3D position of human body joints. The whole skeleton parts are divided for accuracy of the competitive test set.	Randomized decision forests	mocap data	The findings also provide a high correlation between real and given data between classification of the intermediate and the final joint.
Vemulapalli et al. (2014) [13]	Using rotations and translations that use Lie group by explicit modeling, and 3D skeleton geometric relationships are carried out between different body sections.	3D skeleton model, Lie group model	MSR-Action3D, Kinect-Action and Florence3D-Action	Recognition of skeleton-based human behavior strategies in all three datasets, with an accuracy of about 90%
Du et al. (2015) [28]	The human skeleton is divided into five sections and then fed into five sub-networks.	Hierarchical bidirectional and unidirectional RNN	MSR Action 3D, Berkeley MHAD, and HDM05	It was not possible to separate the planned work from just the skeleton joints.
Du et al. (2016) [15]	In order to identify the behavior, hierarchical RNN is used to derive the representation of temporal dynamic of skeleton orders.	RNN and LSTM	ChaLearn, HDM05, Berkeley MHAD and MSR-Action3D	The appearance and scene information was lacking in the proposed work, which discriminates significantly in consideration of action.
Majd and Safabakhsh, (2019) [29]	Using convolution and correlation operators that are able to extract spatial and motion features and their temporal dependencies, the new unit was formed.	Correlational Convolutional LSTM	UCF101 and HMDB51	From correlation details, it was intended to analyze different unit architectures. For good performance, the network architecture could be changed.
Yan et al. (2019) [20]	Recognizing human activity using CSI-based Wi-fi to analyze common human behaviors.	Channel State Information (CSI), Adaptive Activity Cutting Algorithm (AACA) and Extreme Learning Machine (ELM)	-	In the proposed work doppler shift, correlation values are used in WiAct for ELM. Results of the ELM show high accuracy of 94.20%.

(Continued)

TABLE 8.2 (Continued)
Experimental Result Acquired Using Different Models

Paper	Highlight	Method	Dataset	Outcomes and Future Work
Zhang et al. (2019) [19]	By using a graph edge convolution network for action recognition, the proposed work represents each edge by integrating its neighboring edges.	Graph Edge Convolutional Neural Networks, Order-Level Hybrid Model and Body-Part-Level Hybrid Model	NTU-RGB+D Kinetics	It captures the correlation and dependencies between human limbs.
Nadeem et al. (2020) [35]	Using ANN to monitor and identify human behavior based on recognition of body parts.	Linear discriminant and ANN	Weizmann Human action dataset and KTH-dataset	As these acts are special and multidimensional functions, the confusion matrix for KTH-dataset offers 100% accuracy for running with the ball, dribbling, and standing. In the meanwhile, take a pass, shoot, run, and pass. The mean accuracy of identification is 87.57%.
Zhao and Jin (2020) [34]	The proposed work uses Bi-LSTM model of p-non-local and Fusion Key Less Attention to achieve human behavior.	Bi-LSTM No-local and CNN model	HMDB51	The p-non-local block reduces the computational complexity of long-distance dependencies, ensuring efficiency.
Xuemin et al. (2020) [36]	The built-in twin-based digital assembly, commissioning total factor information model for aerospace style, high-precision, electro-hydraulic, servo valve.	Assembly Commission method, Pareto optimal method, and Digital twin model	-	Assembly accuracy of the military product was lower than the micron level and there is no deep discussion about upstream and downstream.

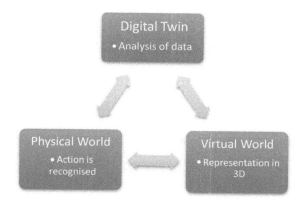

FIGURE 8.3 Digital twin model.

FIGURE 8.4 Architecture of human Action recognition.

of a city, then the working will be the same as at the original site. It will monitor by taking all the necessary data and analyze its future necessities or capabilities. If any kind of problem is there, it will be able to predict for the coming future. The digital twin layer will inform beforehand and will also suggest the best way to solve it. Right now, it asks for suggestions from humans but in the coming future, it might be able to make its decision without human implication. The first step for this goal is to be able to find the activity performed in its surrounding.

Human Action Recognition is categorized into various steps. Figure 8.4 shows the architecture of human action recognition.

The first step is data acquisition, in which data is first captured and then pre-processed. After pre-processing, the data is segmented. The required features are extracted from segmented data. Then dimension reduction takes place and at last, the data is classified and we get the desired information.

8.4.1 Data Acquisition

Recognition of human behavior (Figure 8.5) may be sensor-based or vision-based. Knowledge of human behavior from a data perspective is limited to developments that involve RGB data, depth data, sensor data, or skeleton data [37].

You can collect sensor information from various sensors, such as accelerometers, motion sensors, biosensors, gyroscopes, pressure sensors, proximity sensors, RFID, etc. They can be mounted to various items or can be used for the setting, as sensing

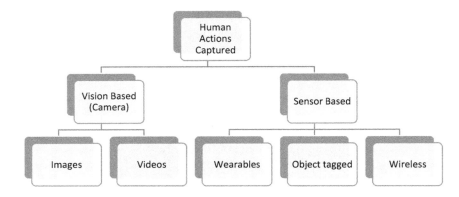

FIGURE 8.5   Data acquisition.

devices or installed. These methods have solved a variety of recognition problems with RGB cameras or video or Wi-Fi to a specific extent and demonstrated good recognition efficiency. Different computer vision techniques have been developed which can process and analyze the data from the camera or sensor and can recognize activities automatically. The activities can be recognized as an action that is based on gesture recognition, posture recognition, behavior recognition, etc., or can be based on human-object interconnection. It can also be motion-based, like tracking, motion identification, or people counting.

8.4.2   PRE-PROCESSING

Data that we gather need to be pre-processed before further analysis. As data can be incomplete, noisy, inconsistent, and may not have the quality which can be useful, the data mining process is required which performs the following tasks as below:

- **Data cleaning**: It is the process of removing or modifying incorrect, corrupt, duplicate, or incomplete data.
- **Data integration**: It means combining information from various sources into a unified view. It efficiently manages data and makes it available to those who need it.
- **Data transformation**: It converts data from one format to another.
- **Data reduction**: It reduces the amount of capacity required to store data and increases storage efficiency at reduced cost.
- **Data discretization**: It converts the continuous data that attributes values to a discrete counterpart.

8.4.3   SEGMENTATION

After the raw data is being pre-processed, it is required to be segmented. Image segmentation provides a powerful semantic description of video imagery, essential

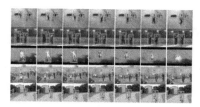

FIGURE 8.6 Data segmentation [38].

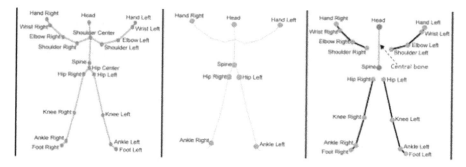

FIGURE 8.7 Data extraction [39].

in image understanding and efficient manipulation of image data. In particular, segmentation based on image motion defines regions undergoing alike motion, which allows the image coding system to effectively show video order. Figure 8.6 shows data segmentation from the data from a different location.

8.4.4 EXTRACTING ATTRIBUTE

Attribute extracting is to automatically extract features from signals or images by creating new features from the existing ones. The dimensions by which an initial collection of raw data is reduced to more manageable groups for processing are reduced in this process. Much of the details contained in the original set of features should then be able to give information about a new reduced set of features. A summarized version of the original features can be created from a variation of original features. Figure 8.7 shows data extraction, where each part is extracted from the data used.

8.4.5 DIMENSIONALITY REDUCTION

Dimensionality reduction refers to techniques for reducing the number of input variables in training data. When dealing with high dimensional data, it is often useful to reduce dimensionality by projecting the data to a lower-dimensional subspace which captures the essence of data. Avoiding overfitting is a major motivation for performing dimensionality reduction.

8.4.6 CLASSIFICATION

Classification is a supervised approach (as shown in Figure 8.8) wherein the computer program learns from the data it receives and makes discoveries or classifications. It is a method by which a given collection of information is classified that can be carried out on structured or unstructured data. The process starts with the class of given data points being predicted. The groups are also referred to as divisions, marks, or goals. The predictive model of labeling is its process of approximating the mapping function to distinct output variables through input variables. The primary objective is to identify the group in which the new data will fall.

Although, two main problems, including the real case are in terms of interconnection recognition and action identification. Interconnection represents steps that include two persons or actions between persons and objects. Action identification includes identification of location, wherein the action happens in time and space from image order detail, which is not separated. Deep learning models are the type of machines that will acquire a ranking of specifications by constructing higher-level specifications from lower specifications [40]. The convolutional neural networks (CNNs) are deep models in which trainable filters and native neighborhood pooling operations are applied on raw input images, which results in a ranking of highly complex specifications. It is proved that CNNs when trained with proper regularization can do an extremely good presentation on visual perception role. Also, CNN's are indifferent to variations such as pose, lighting, and surrounding clutter.

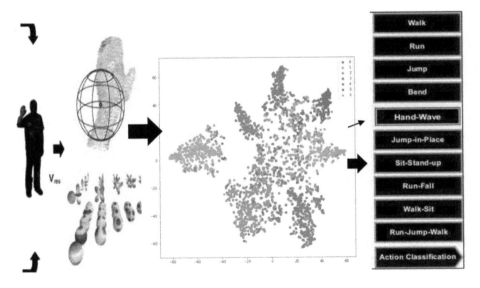

FIGURE 8.8 Data classification.

8.5 CHALLENGES FACED IN HUMAN ACTION RECOGNITION IN DIGITAL TWIN ENVIRONMENT

The identification of actions taken by humans is increasing and is applied in several companies. Although it has positive implications, there are some drawbacks also, as given below:

- Background jumble

    The background jumble suggests that there are human beings present in the picture and it may be difficult for anyone to concentrate on a single person. Recognition of action is itself a difficult issue and the inclusion of background jumble makes the issue more difficult as it would be difficult to distinguish the person in the video from the freely moving individuals. The main area of the picture is obscured by the shadow.

- Partial obstruction

    Obstruction of a local region of the body with objects such as sunglasses, scarves, hands, watches, etc. is generally called partial occlusions. The obstruction generally has to be less than 50% to be scrutinized as a partial occlusion.

- Scale change

    One of the major issues in dissipated environments is that of scale; the action of a human captured at large distances is considerably harder to recognize than the same human's action at small distances. This problem is very common in image classification.

- Perspective

    Perspective variation occurs when the same image is viewed from different angles i.e., rotated or oriented in multi-dimensions, concerning how the image or video is captured. No matter the angle in which we capture the image of human action, whether it's a still or an act of drinking, jumping, etc.

- Illumination

    The image or video captured of a human being might be taken at different illumination or brightness level, which makes it difficult to identify. Our image classification system must be able to grasp the dissimilarity in illumination. So, when we give any image of the same human's action with different brightness levels to our image classification system, it should be able to identify the same action.

- Similarities in inter and intra class

    Having variance is intra-class, that is, in a picture of the same class, while the changes between images that have various class labels are determined by inter-class variations. In each sample problem, since there is only an image per individual, all the variations are inter-class variations.

Because of these challenges, recognition becomes difficult. When these data are fed in the digital twin layer, it will not be able to predict the correct result, which will affect the future planning of that physical world.

8.6 HUMAN FEATURE RECOGNITION BASED ANALYSIS FOR DIGITAL TWIN USING CNN MODEL

It is important to extract features of human behavior in digital twin environment and to select all the significant human characteristics that enhance efficiency of the model of machine learning or deep learning. For these neural networks, the disappearing and exploding gradient is a common issue altogether. The backpropagation algorithm is linked to this problem. We will address various forms of neural networks used below, in the recognition of human behavior.

8.6.1 ROLE OF ANN IN HUMAN ACTION RECOGNITION

As inputs are interpreted only in forward direction, ANN is referred to as a Feed-Forward Neural network. As a gaggle of multiple neurons in each sheet, ANN is taken into account. As shown in Figure 8.9, ANN has three layers: the input layer, the secret layer, and the output layer. The input layer accepts inputs, the hidden layer processes the inputs, and the result is then generated by the output layer. Each layer attempts to classify those weights. The Artificial Neural Network is good at learning any nonlinear function that enables the network to understand complex input-output relationship. The key move is to transform a two-dimensional image into a one-dimensional image, when solving a picture classification problem using ANN.

The use of ANN has two drawbacks. First, the number of trainable parameters increase dramatically, with an increase in the image size, and second, ANN lacks a picture's spatial features that ask for the pixels in an image to be organized.

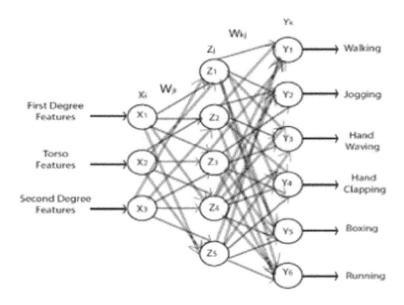

FIGURE 8.9 ANN structure [35].

8.6.2 Role of CNN in Human Action Recognition

Convolutional neural networks (CNN) are all the craze within the deep learning community recently. These CNN models are getting used across different applications and domains. They are especially prevalent in image and video processing.

The convolutional layer is composed of a set of convolutional kernels, where each neuron acts as a kernel. Although, if the kernel is symmetric, the convolution operation becomes a correlation operation. Convolutional kernel works by dividing the image into small slices, commonly known as receptive fields. The division of an image into small blocks helps in extracting feature motifs. Kernel convolves with the images using a specific set of weights by multiplying its elements with the corresponding elements of the receptive field.

CNN can be built with several layers. Each neuron is connected with every other neuron of the next layer. Every layer gets its input from the output of the previous layer. There might be convolution layers, pooling layers, and fully connected layers, related to a common activation function. At last, there is often a SoftMax layer for classification. So, CNN is often imagined as a series of layers. Figure 8.10 shows the CNN architecture.

CNN learns to filter automatically, that is, without directly stating it. These filters assist in extracting from the input data of human behavior with the right and appropriate features. From an image, CNN captures the spatial features. Spatial characteristics are the groupings of pixels that accurately identify the link between the behavior of human beings and their surroundings. The location of a human being in a frame is often correctly identified by its association with another environment. The conception of parameter sharing is also followed by CNN. One filter is implemented as a function map through various parts of an input that needs to be supplied.

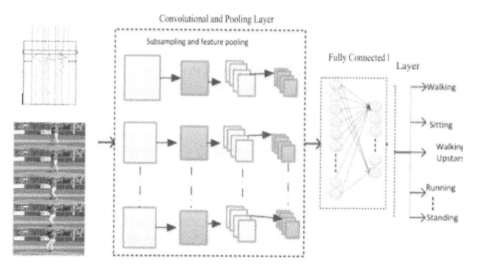

FIGURE 8.10   CNN Structure [41].

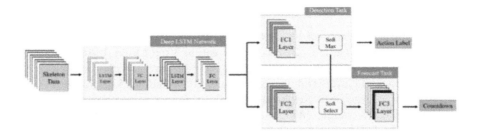

FIGURE 8.11   RNN Structure [42].

8.6.3   ROLE OF RNN IN HUMAN ACTION RECOGNITION

RNN captures dependency within the image while making predictions. It can also process both single data points, such as images, and sequences of data, such as speech, video, human activity, etc. RNNs share the parameters across different time pace shown in Figure 8.11. It is often popularly referred to as Parameter Sharing. It leads to fewer parameters to train and reduce computational cost. D-RNNs are RNNs, with a huge time pace that also suffer from vanishing and exploding gradient problem.

The RNN type of Long Short-Term Memory (LSTM) consists of a cell and three regulatory agencies in each unit, namely input and output gates, and a forget gate. The values of arbitrary time intervals are remembered by each cell, while the gates control the flow of information. These networks are suitable for time-series-based data classification, processing, and prediction. In learning, processing, and classifying such types of data, LSTMs have proven to excel. It's not only a series of layers but also has stacked up time-pace layers. It may be imagined as a mesh of layers, horizontally as well as vertically. The main difference between CNN and RNN is that RNN has time steps. In other words, it can process the live streams. Below (Table 8.3) is the comparative representation of ANN, CNN, and RNN.

8.7   APPLICATION AREA OF DIGITAL TWIN WITH HUMAN ACTION RECOGNITION

Nowadays, computers are omnipresent and serve as facilitation of our daily work and life.

8.7.1   MONITORING DAILY ACTIVITIES

Earlier, human actions were evaluated by human operators, for example, security and surveillance processes or monitoring the health status of a patient. When the number of camera viewing devices and technological monitoring devices increased, so did the number. The task of managing and controlling operators is demanding and expensive, as a ceaseless operation is now necessary. In the case of home treatment, certain activities will also not be financially viable for staff deployment. So, monitoring daily activities is now done automatically with the advancement of technology,

TABLE 8.3

Comparison between ANN, CNN, and RNN in Human Recognition [43]

	ANN	CNN	RNN
Data Type	Tabular data, sensor-based data	Mainly images but can be used for the sequential image too.	Works better on a series of image (video)
Performance	Less powerful	More powerful.	Less than CNN but better than ANN
Main advantages	Fault forbearance can work with insufficient data	Shares weight and have a high-performance rate in activity recognition from images	Predicts time orders and remembers every information
Disadvantages	Hardware dependent and unaccepted behavior of the network.	A large amount of training data is required and does not encode an individual's position and inclination.	A situation like a gradient vanishing and exploding gradient occurs.

using deep learning and machine learning techniques. Most of the actions are identified correctly, using these algorithms but not all are correct. These actions, when used in the digital twin environment, can help to manage the disaster, reduce consumption of resources, and many more. For this purpose, many researchers are working to get the best performance for improving the living of the society. It can be used for security purposes, for checking on in-home patients, to take care of family members, in schools for looking after the children and managing works, and many more.

8.7.2 DIGITAL TWIN IN HEALTHCARE

As a digital twin provides real-time analysis with equipment and other physical assets, they can revolutionize healthcare operations. A digital twin of the patient organ with recognition of the doctor's action can allow the surgeon to practice procedure and be prepared, before the actual operation. Also, the need for elderly care is growing rapidly as the Baby Boomer generation is getting retired. A major goal of current research in human activity, monitoring and digital twin analysis, is to develop new technologies and applications for elderly care. Those applications could help prevent harm, such as, by detecting dangerous situations for older people. Changes in the architecture of the smartphone identifies if the user falls down. Activity recognition and monitor sensors could help elders in a proactive way, such as life routine reminders (e.g., taking medicine), living activity monitoring for a remote robotic assistance, among others. Youth care is another field that benefits from the activity recognition research. Applications include monitoring infants' sleeping status and predicting their demands for food or other stuff.

8.7.3 IMPACT IN INDUSTRIES

In the vision of Industry 4.0, tasks, and responsibilities among human employees and robots are shared [44]. Automated robots can't fully replace manual labor in the foreseeable future. It is difficult for robots to emulate the rational aptitude of humans [45,46]. But, because of the advancements in sensor technology and data processing, IT-supported approaches for automated activity recognition and assessment are gaining significance. Tarik Uzunovic et al. [47] combined human activity recognition and task-based control, to establish successful collaboration. Wherein, robots are supporting humans for many tasks. Examples of human-machine interconnection are Multi-Touch Technology [48], Smartwatch [49], Deco Exoskeleton, and many more. Recently, digital twin gave a new concept to the industry. Now we have twins of each product that analyze the working and provide us a future prediction. For new products or factories [50], it is first tested in a virtual environment, to check the product's capability and changes are made accordingly before launching in the market. This reduces most of the risk and helps the industry to work efficiently and saves time and money [51,52].

8.7.4 LARGE-SCALE GOVERNMENT IMPLEMENTATION TO MANAGE DISASTER AND ANOMALOUS ACTIVITY

By interpreting and understanding human activity, we can recognize and predict the occurrence of disaster and crime and help the government, police, or other agencies to react immediately. A digital twin environment of a city is created for the analysis [24]. It is predominant to identify human factors that impact successful implementation of port security measures. To reliably associate data for a particular person, a successful and efficient identification scheme must be established and maintained. This will allow the identification of disaster beforehand, according to the past decade's data provided to the machine. Enormous papers are on human activity recognition in video and image orders. One of them is a framework for recognizing human activity from the environment to identify anomalous activities [53,54]. There is a need to find people, using a laser range-finder in an outdoor surrounding and automatically differentiate tracks of individuals into sub-tracks, representing different activities [55]. Similarly, there is a need to find an abnormal task in smart homes for the families like children, elderly, and disabled people [56], and deep learning architecture for recognizing activities from groups that are anonymous using context and motion information [57].

8.8 CONCLUSION

With the advancement of digital twin, human action recognition remains a significant issue in computer vision. Human action recognition has improved tremendously over the decades that helps in perceiving data from the surrounding into the digital twin layer. Now we have different data sets that have increased the accuracy rate, also due to the use of different deep learning techniques, used for data extraction and classification. But the problem still arises in identifying the action accurately, when confronting sensible landscapes, notwithstanding the characteristic intraclass variety and interclass similitude issue, which affects prediction of Digital twin. This paper presents some of the techniques and problems faced, which need to be improvised further, so to improve the overall performance. As for the future direction, a more accurate algorithm for action recognition is also needed, that when it is applied on different data set, it shows the same accuracy level, with more efficient representation, and real-time operations still remaining open problems. This will allow the digital twin to make a better prediction to some extent.

REFERENCES

1. Howard, George S., William H. Youngs, and Ann M. Siatczynski. "A research strategy for studying telic human behavior," *The Journal of Mind and Behavior* 10.4 (1989): 393–411. Accessed October 23, 2020. http://www.jstor.org/stable/43853474.
2. Fujiyoshi, Hironobu, Alan J. Lipton, and Takeo Kanade. "Real-time human motion analysis by image skeletonization," *IEICE Transactions on Information and Systems* 87.1 (2004): 113–120.
3. Kemp, Martin. *Leonardo da Vinci: The Marvellous Works of Nature and Man.* Oxford University Press, 2007. New York
4. Provencher, Matthew, and William Abdu. "Historical perspective: Giovanni Alfonso Borelli: "Father of Spinal Biomechanics"," *Spine* 25.1 (2000): 131.
5. Verfaillie, Karl. "Perceiving human locomotion: Priming effects in direction discrimination," *Brain and Cognition* 44.2 (2000): 192–213.
6. Fuller, Peter W.W. "Life and work of Eadweard Muybridge." *23rd International Congress on High-Speed Photography and Photonics.* Vol. 3516. International Society for Optics and Photonics, 1999.
7. Johansson, Gunnar. *Visual Vector Analysis and the Optic Sphere Theory.* Lawrence Erlbaum Assoc., Hillsdale, 1994.
8. Gavrila, Dariu M. "The visual analysis of human movement: A survey," *Computer Vision and Image Understanding* 73.1 (1999): 82–98.
9. Ioffe, Sergey, and David Forsyth. "Finding people by sampling." *Proceedings of the Seventh IEEE International Conference on Computer Vision,* IEEE, 1999, vol. 2.
10. Lee, Lae Kyoung, Sungshin Kim, Young-Kiu Choi, and Man Hyung Lee. "Recognition of hand gesture to human-computer interconnection," *2000 26th Annual Conference of the IEEE Industrial Electronics Society. IECON 2000. 2000 IEEE International Conference on Industrial Electronics, Control and Instrumentation. 21st Century Technologies,* Nagoya, Japan, 2000, vol. 3, pp. 2117–2122, doi: 10.1109/IECON.2000.972603.
11. Shimohata, Yasuyuki, and Nobuyuki Otsu. "Real-time and simultaneous recognition of multiple moving objects using cubic higher-order local auto-correlation," *2008 IEEE Southwest Symposium on Image Analysis and Interpretation,* IEEE, 2008.
12. Delaitre, Vincent, Ivan Laptev, and Josef Sivic. "Recognizing human actions in still images: A study of bag-of-features and part-based representations." *BMVC 2010-21st British Machine Vision Conference.* 2010.
13. Vemulapalli, Raviteja, Felipe Arrate, and Rama Chellappa. "Human action recognition by representing 3D skeletons as points in a lie group," *Proceedings of the IEEE Conference on Computer Vision and Pattern Recognition,* 2014.
14. Chen, Chen, Roozbeh Jafari, and Nasser Kehtarnavaz. "UTD-MHAD: A multimodal dataset for human action recognition utilizing a depth camera and a wearable inertial sensor," *2015 IEEE International Conference on Image Processing (ICIP),* IEEE, 2015.
15. Du, Yong, Yun Fu, and Liang Wang. "Representation learning of temporal dynamics for skeleton-based action recognition," *IEEE Transactions on Image Processing* 25.7 (2016): 3010–3022.
16. Uzunovic, Tarik, Edin Golubovic, Zlatan Tucakovic, Yasin Acikmese, and Asif Sabanovic. "Task-based control and human activity recognition for human-robot

collaboration," In *IECON 2018–44th Annual Conference of the IEEE Industrial Electronics Society*. IEEE, 2018, pp. 5110–5115.

17. Zeeshan Khan, Muhammad, Muhammad A. Hassan, Ammarah Farooq, and Muhammad Usman Ghanni Khan. "Deep CNN based data-driven recognition of cricket batting shots," *2018 International Conference on Applied and Engineering Mathematics (ICAEM)*, Taxila, 2018, pp. 67–71, doi: 10.1109/ICAEM.2018.8536277.

18. Gao, Xuesong, Keqiu Li, Yu Zhang, Qiguang Miao, Lijie Sheng, Jun Xie, and Jinfu Xu. "3D skeleton-based video action recognition by graph convolution network," In 2019 IEEE International Conference on Smart Internet of Things (SmartIoT). IEEE, 2019, pp. 500–501.

19. Zhang, Xikun, Chang Xu, Xinmei Tian, and Dacheng Tao. "Graph edge convolutional neural networks for skeleton-based action recognition," *IEEE Transactions on Neural Networks and Learning Systems* 31.8 (August 2020): 3047–3060, doi: 10.1109/TNNLS.2019.2935173.

20. Yan, Huan, Yong Zhang, Yujie Wang, and Kangle Xu. "WiAct: A passive Wi-Fi-based human activity recognition system," *IEEE Sensors Journal* 20.1 (2020): 296–305, doi: 10.1109/JSEN.2019.2938245.

21. Yilmaz, Abdullah Asim, Mehmet Serdar Guzel, Erkan Bostanci, and Iman Askerzade. "A novel action recognition framework based on deep-learning and genetic algorithms," *IEEE Access* 8 (2020): 100631–100644, doi: 10.1109/ACCESS.2020.2997962.

22. Wu, Xiao, and Junsong Yuan. "Multipath event-based network for low-power human action recognition," *2020 IEEE 6th World Forum on the Internet of Things (WF-IoT)*, New Orleans, LA, USA, 2020, pp. 1–5, doi: 10.1109/WF-IoT48130.2020.9221355.

23. Peng, Cheng, Haozhi Huang, Ah-Chung Tsoi, Sio-Long Lo, Yun Liu, and Zi-yi Yang. "Motion boundary emphasized optical flow method for human action recognition," *IET Computer Vision* 14.6 (2020): 378–390, doi: 10.1049/iet-cvi.2018.5556.

24. Fan, Chao, Cheng Zhang, Alex Yahja, and Ali Mostafavi. "Disaster city digital twin: A vision for integrating artificial and human intelligence for disaster management," *International Journal of Information Management* 56 (2021): 102049.

25. Nikolaos Nikolakis, Kosmas Alexopoulos, Evangelos Xanthakis, and George Chryssolouris. "The digital twin implementation for linking the virtual representation of human-based production tasks to their physical counterpart in the factory-floor," *International Journal of Computer Integrated Manufacturing* 32.1 (2019): 1–12, doi: 10.1080/0951192X.2018.1529430.

26. Guanghui Zhou, Chao Zhang, Zhi Li, Kai Ding, and Chuang Wang, "Knowledge-driven digital twin manufacturing cell towards intelligent manufacturing," *International Journal of Production Research* (2019), doi: 10.1080/00207543.2019.1607978.

27. Shotton, Jamie, Toby Sharp, Alex Kipman, Andrew Fitzgibbon, Mark Finocchio, Andrew Blake, and Richard Moore. "Real-time human pose recognition in parts from single depth images," *Communications of the ACM* 56.1 (2013): 116–124.

28. Du, Yong, Wei Wang, and Liang Wang. "Hierarchical recurrent neural network for skeleton based action recognition," *Proceedings of the IEEE Conference on Computer Vision and Pattern Recognition*, 2015, pp. 1110–1118.

29. Majd, Mahshid, and Reza Safabakhsh. "Correlational convolutional LSTM for human action recognition," *Neurocomputing* 396 (2020): 224–229.

30. Sahoo, Suraj Prakash, R. Silambarasi, and Samit Ari. "Fusion of histogram-based features for Human Action Recognition," *2019 5th International Conference on Advanced Computing & Communication Systems (ICACCS)*, Coimbatore, India, 2019, pp. 1012–1016, doi: 10.1109/ICACCS.2019.8728473.

31. Park, Sungjoo and Dongchil Kim. "Study on 3D action recognition based on deep neural network," *2019 International Conference on Electronics, Information, and Communication (ICEIC)*, Auckland, New Zealand, 2019, pp. 1–3, doi: 10.23919/ELINFOCOM.2019.8706490.

32. Gao, Xuesong, Keqiu Li, Yu Zhang, Qiguang Miao, Lijie Sheng, Jun Xie, and Jinfu Xu. "3D skeleton-based video action recognition by graph convolution network," In *2019 IEEE International Conference on Smart Internet of Things (SmartIoT)*, Tianjin, China, 2019, pp. 500–501, doi: 10.1109/SmartIoT.2019.00093.

33. Dröder, Klaus, Paul Bobka, Tomas Germann, Felix Gabriel, and Franz Dietrich. "A machine learning-enhanced digital twin approach for human-robot-collaboration," *Procedia CIRP* 76 (2018): 187–192.

34. Zhao, Han, and Xinyu Jin. "Human action recognition based on improved fusion attention CNN and RNN," *2020 5th International Conference on Computational Intelligence and Applications (ICCIA)*, Beijing, China, 2020, pp. 108–112, doi: 10.1109/ICCIA49625.2020.00028.

35. Nadeem, Amir, Ahmad Jalal, and Kibum Kim. "Human actions tracking and recognition based on body parts identification via artificial neural network," *2020 3rd International Conference on Advancements in Computational Sciences (ICACS)*, Lahore, Pakistan, 2020, pp. 1–6, doi: 10.1109/ICACS47775.2020.9055951.

36. Sun, Xuemin, Jinsong Bao, Jie Li, Yiming Zhang, Shimin Liu, and Bin Zhou. "A digital twin-driven approach for the assembly-commissioning of high precision products," *Robotics and Computer-Integrated Manufacturing* 61 (2020): 101839.

37. Majumder, Sharmin, and Nasser Kehtarnavaz. "Vision and inertial sensing fusion for human action recognition: A review," *IEEE Sensors Journal* 21.3 (2021): 2454–2467.

38. Yu, Gang, and Junsong Yuan. "Fast action proposals for human action identification and search," In *Proceedings of the IEEE Conference on Computer Vision and Pattern Recognition*, 2015, pp. 1302–1311.

39. Warchoł, Dawid, and Tomasz Kapuściński. "Human action recognition using bone pair descriptor and distance descriptor," *Symmetry*, 12.10 (2020): 1580.

40. Koohzadi, Maryam, and Nasrollah Moghadam Charkari. "Survey on deep learning methods in human action recognition," *IET Computer Vision* 11.8 (2017): 623–632.

41. Nweke, Henry Friday, Ying Wah Teh, Mohammed Ali Al-Garadi, and Uzoma Rita Alo. "Deep learning algorithms for human activity recognition using mobile and wearable sensor networks: State of the art and research challenges," *Expert Systems with Applications* 105 (2018): 233–261.

42. Liu, Chunhui, Yanghao Li, Yueyu Hu, and Jiaying Liu. "Online action identification and forecast via multitask deep recurrent neural networks," In *2017 IEEE International Conference on Acoustics, Speech and Signal Processing (ICASSP)*, IEEE, 2017 March, pp. 1702–1706.

43. Abhishek Gupta, "Difference between ANN, CNN and RNN," https://www.geeksforgeeks.org/difference-between-ann-cnn-and-rnn, last updated: 17 July 2020.

44. Reining, Christopher, Friedrich Niemann, Fernando Moya Rueda, Gernot A. Fink, Michael ten Hompel,"Human activity recognition for production and logistics—A systematic literature review," *Information* 10.8 (2019): 245.

45. Breazeal, Cynthia, and Brian Scassellati. "Robots that imitate humans," *Trends in Cognitive Sciences* 6.11 (2002): 481–487.

46. Langley, Pat, John E. Laird, and Seth Rogers. "Cognitive architectures: Research issues and challenges," *Cognitive Systems Research* 10.2 (2009): 141–160.

47. Uzunovic, Tarik, Edin Golubovic, Zlatan Tucakovic, Yasin Acikmese, and Asif Sabanovic. "Task-based control and human activity recognition for human-robot collaboration," In IECON 2018-44th Annual Conference of the IEEE Industrial Electronics Society. IEEE, 2018, pp. 5110–5115.

48. Westerman, Wayne, John G. Elias, and Alan Hedge. "Multi-touch: A new tactile 2-D gesture interface for human-computer interconnection," *Proceedings of the Human Factors and Ergonomics Society Annual Meeting*, SAGE Publications, Sage, CA: Los Angeles, CA, 2001, vol. 45. no. 6.

49. Chen, Xiang'Anthony, Tovi Grossman, Daniel J. Wigdor, and George Fitzmaurice. "Duet: Exploring joint interconnections on a smart phone and a smart watch," In Proceedings of the SIGCHI Conference on Human Factors in Computing Systems. 2014, pp. 159–168.

50. Yildiz, Emre, Charles Møller, and Arne Bilberg. "Virtual factory: Digital twin based integrated factory simulations," *Procedia CIRP* 93: 216–221.

51. Tao, Fei, He Zhang, Ang Liu, and A. Y. Nee. "Digital twin in industry: State-of-the-art," *IEEE Transactions on Industrial Informatics* 15.4 (2018): 2405–2415.

52. Zhou, Guanghui, Chao Zhang, Zhi Li, Kai Ding, and Chuang Wang. "Knowledge-driven digital twin manufacturing cell towards intelligent manufacturing," *International Journal of Production Research* 58.4 (2020): 1034–1051.

53. Xu, Xin, Jinshan Tang, Xiaolong Zhang, Xiaoming Liu, Hong Zhang, and Yimin Qiu. "Exploring techniques for vision based human activity recognition: Methods, systems, and evaluation," *Sensors* 13.2 (2013): 1635–1650.

54. Vishwakarma, Sarvesh, and Agrawal, Anupam. "A survey on activity recognition and behavior understanding in video surveillance," *Visual Computer* 29 (2013): 983–1009.

55. Panangadan, Anand, Maja Mataric, and Gaurav Sukhatme. "Identifying anomalous human interconnections using laser range-finders," *2004 IEEE/RSJ International Conference on Intelligent Robots and Systems (IROS) (IEEE Cat. No. 04CH37566)*, IEEE, 2004.

56. Saqaeeyan, Sasan, Hamid Haj Seyyed javadi, and Hossein Amirkhani. "Anomaly identification in smart homes using Bayesian networks." *KSII Transactions on Internet & Information Systems* 14.4 (2020): 1796–1816.

57. Borja-Borja, Luis Felipe, Jorge Azorín-López, and Marcelo Saval-Calvo. "A deep learning architecture for recognizing abnormal activities of groups using context and motion information," *International Workshop on Soft Computing Models in Industrial and Environmental Applications*, Springer, Cham, 2020, pp. 760–769.

9 eVote – A Decentralized Voting Platform

Vanita Jain, Akanshu Raj, Abhishek Tanwar, and Mridul Khurana
Bharati Vidyapeeth's College of Engineering

CONTENTS

9.1   INTRODUCTION

Democracy [1] is a Greek word that means rule by the people. Democracy-based government is the best type of government for what it's worth, of the individuals, by the individuals. and for the individuals. In a democratic country, people are allowed to choose their leaders. The process of selecting the leader is called Elections [2]. A hint of traces can be discovered back in history [3], in old Greece and antiquated Rome, and all through the Medieval time-frame to elections (a type of conversation) that were used to choose rulers. Be that as it may, the present-day type of "political decision" that comprises open appointment of government authorities, didn't develop until the start of the seventeenth century when the thought of delegated government grabbed attention in North America and Europe.

Elections [2,4] have become an indistinguishable part of democratic government. People get to choose their leaders by voting for them. The right to vote [5] is one of the basic rights, and all the qualified individuals must get it.

However, in today's world, we have seen many cases where this right has been violated [6]. There have been many instances where people claim that the vote didn't go to the person or party they wanted. Sometimes, people who haven't voted find that

DOI: 10.1201/9781003132868-9

someone else has voted using their ID's. Many issues are discussed later in this paper, and a solution to overcome all those shortcomings is proposed.

9.2 BALLOT PAPER

Ballot papers [7,8] are used to cast votes in elections. It is a piece of paper used for voting. Each voter has one paper on which they have to write the name of the candidate. But to ensure security, the government started using printed ballots [9]. Voters can cast their votes at polling stations. The voting with the help of ballot paper was first utilized in America [10] in 1629 inside the Massachusetts Bay Colony, to choose a minister for Salem Church.

In the vast majority of nations, the voting forms used to be pre-printed with the names of competitors standing for a post. In countries like the Philippines (until 2007) and Japan, voters used to write the name of the competitor on a voting form paper. Election Officials had to count all the votes. In the case of cheating or dispute, votes were counted again by the officials.

In countries like the Philippines, notwithstanding a tight political race by Election Commission, the vote-buying [11] was done in obscurity, where individuals accumulate a polling form, with a particular measure of cash connected to it.

There were other instances where the unethical means of casting votes continued. For example, in Argentina, there was a secret ballot [12] that was utilized to cast the votes. Many people got voter ballots with a candidate's name written on it. Due to introduction of ballots, many more cases of cheating and monitoring were reported all over the world. Voting form papers were also scrutinized for their plans, there was an issue that occurred during the presidential elections of United States in the year 2000, Butterfly ballot [13] paper was scrutinized and found to be inadequate, as a few voters wound up voting for an applicant whom they did not wish to vote for. This happened because voters got confused due to the design of the ballot paper.

Another strategy for befuddling individuals into casting a ballot for an alternate applicant was to make parties with similar names or symbols that will deceive the voters. Such tactics by people, make the use of ballot paper less secure [14].

Ballot Stuffing [15] is another way by which people can cast multiple votes. Ballot stuffing is also called "ballot-box-stuffing". In an instance that took place in 1883, when the elections for the district of Cook, in Queensland, Australia took place. Numerous individuals had enjoyed the illicit practice of stuffing the ballot [16], which led to several arrests. There was one more instance of ballot stuffing that occurred during the Russian Presidential Election [17], where people were recorded on camera, stuffing the ballot in favor of the President, Vladimir Putin.

In nations like India, it was discovered that goons of political parties capture polling stalls and force voters to vote for their party and in some cases, they cast vote in the name of genuine voters [18].

Earlier votes were counted manually, which can sometimes lead to false recording of the ballot. Counting votes also required some additional investment to tally votes before proclaiming the outcome; Moreover, the individuals who are physically challenged, face difficulty in voting through paper polling form. Considering all the consequences of using ballot papers, Electronic Voting Machines were introduced.

9.3 ELECTRONIC VOTING MACHINES

Electronic voting [19] is a democratic procedure that utilizes machines instead of ballot paper to project votes and also helps to tally the number of votes. Electronic voting technology involves Electronic voting machines [20]. They were legally used in Estonia for the first time. [21].

Electronic Voting Machines ended up being more effective than polling form papers. Machines incorporated different conspicuous highlights that were not present in the ballot paper method; some of its advantages are vote recording, information encryption, complete arrangement of vote input, transmission to workers, and combination and classification of political decision results. Electronic voting paces up the procedure of vote tallying as calculations are done by machines. It also reduces the chances of errors that can occur during the counting of votes and also diminishes the expense of paying staff as there is no need for them to count the votes [22]. The results also get published faster. Unlike in the paper ballot system, voters can't vote more than once. The whole process of electronic voting is done under the supervision of controlling authority. Controlling authority can be the government or any other party that is organizing elections. The efficiency and turnaround time observed with electronic voting machines becomes necessary when they are used for larger populations. The use of electronic voting machines over the years has given confidence to the voters that their votes are having a major impact on the whole democratic system.

In densely populated nations like India, it is essential to have a fast and efficient voting system. Electronic voting machines in India were brought in 1989 [23] by the Election Commission of India in a joint effort with Bharat Electronics Limited and Political Decision Corporation of India Limited. Electronic Voting Machines were first used in India in 1998 for the selection of a few constituencies in state assembly elections. The main reason for introducing Electronic Voting Machines in India is to reduce electoral frauds and to decrease the cost of conducting elections. Electronic Voting Machines replaced ballots throughout the country from 2001 onwards. In 2004, the election was organized entirely through Electronic Voting Machines. In India, Electronic Voting Machines incorporated an essential feature of registering only five votes per minute, thus reducing the risk for vote stuffing that used to happen in the ballot system. In countries that have a significantly large uneducated or illiterate population, electronic voting technology ensures that these groups also participate and their votes are properly counted.

As the voting system became complex and started introducing more software, different methods of electoral fraud increased. A few people tested the utilization of electronic voting systems, and they formed a hypothetical perspective, contending that people are most certainly not prepared for confirming activities happening inside an electronic machine, thus there are chances of manipulation within the machines. Since individuals cannot confirm these activities, thus there is an argument that tasks performed by electronic voting machines can't be trusted.

Numerous issues [23] and thoughts regarding the insecure nature of the EVMs have been reported, one such example is, utilizing a similar default secret key. Many cases of machines making conflicting and eccentric errors were reported [24].

There is likewise no assurance that outcomes are gathered and revealed precisely. There has been contention, especially in the United States, that e-casting a ballot, particularly through DRE (Direct Recording Electronic) [25] could lead to an increase in the number of fakes and may not be completely auditable [26]. Several countries like the Netherlands, Ireland, Germany, and the United Kingdom have canceled the electronic voting system due to reliability issues [27], whereas in countries like India, electronic voting is still in practice.

9.4   ONLINE VOTING SYSTEMS

The continuous issue faced in elections is information control, security, and transparency [28]. With the improvement of innovation, the utilization of new technologies in defeating the issues that happen becomes significant, just as the complexities of the assortment procedure get increased. Security is consistently the greatest worry when it comes to online voting [29,30]. To achieve such a secure system, we can opt for Blockchain [31]. Blockchain is one arrangement that can be utilized to decrease the issues that happen in voting [32].

A straightforward political election can be sorted out, utilizing blockchain innovation. A Blockchain is a time-stepped arrangement of changeless records of information. It is nearly impossible to manipulate information stored in blocks of the blockchain as each block is bounded with the other, utilizing cryptography standards.

What gives blockchain an edge over other technologies is the way the data is recorded. Blockchain is a better approach for reporting information on the web, and the information stored in it is open for anybody and everybody to see. Blockchain is implemented with the help of smart contracts [33], and can be deployed on platforms like Bitcoin and Ethereum [34,35].

There is one more technology that will help in making the elections completely online and this technology is the Inter Planetary File System (IPFS) [36]. It is a conventional and distributed system for putting away and sharing information in a dispersed record framework. It utilizes content, tending to remarkably distinguish each record in a worldwide name and -space. associating all processing gadgets together. Records can only be accessed using a unique address path [37–39]. Every file/document uploaded on IPFS returns a hash value which is unique for every unique document. It means that no two distinct files can have identical hash values.

9.5   PROPOSED SYSTEM AND WORKING

We have proposed a system that combines two states of art technologies, IPFS and Blockchain, which are merged with an interactive, user-friendly interface, created using ReactJS. The proposed system is more secure and transparent than the traditional methods that are in use. We have created a smart contract that uses the features of blockchain, and IPFS is used as a platform to store the identity of the users (eligible voters) in a decentralized manner, which can only be accessed by the entity who has hash key to that document. The working of the whole system is broken into individual components and is explained in great detail.

9.6 THE SMART CONTRACT

The Smart Contract that we have built is deployed on Ethereum Blockchain. For testing purposes, we have deployed a contract on the Rinkeby network and used an injected web3 environment.

The contract is deployed by the authority responsible for conducting the elections. Figure 9.1 shows deployment page of the website. The authority enters its address and then the required description. In description, the authority then mentions the purpose of deploying the contract.

Once the contract is deployed, we get the address of the deployed contract. From Figure 9.2 we can see that we have to add identities of candidates/parties in the contract. The parties/candidates can be added only by the authority that has deployed the contract.

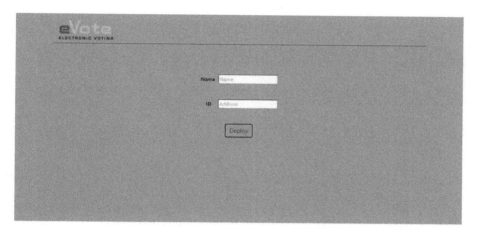

FIGURE 9.1 Smart contract deployment page.

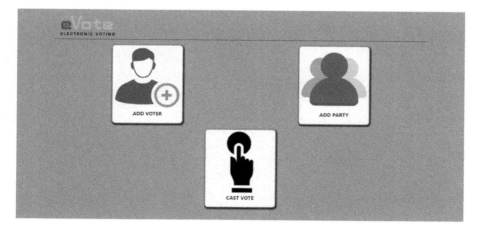

FIGURE 9.2 Page to take required action.

Figure 9.3 shows the page to add parties/candidates. To add parties/candidates, the authority has to add name and description of the candidate/party. Users can access the details of all the parties which includes party description and vote count of that party.

Figure 9.4 shows the add voter page on the website. Eligible voters have to register in order to be validated by first and second controlling authority in order to get their names on the voter list. The controlling authorities will now work to check if the voter is a genuine person and not a fraud. The voter will be then added to the voter list. The controlling officials validate the voters after verifying the details provided by the voter which can only be called by the controlling authorities, this function considers address of the voter as an argument.

Initially, the first controlling official views the details of a requested voter and if details are correct, first controlling official calls the function with the address of corresponding voter and validates that voter.

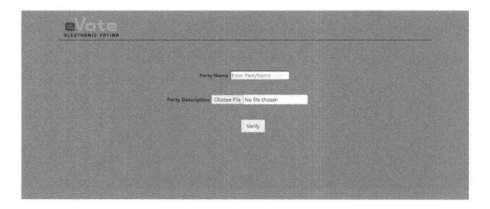

FIGURE 9.3 Web Page to add parties.

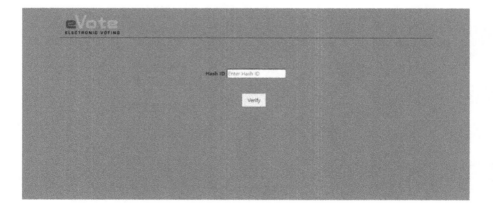

FIGURE 9.4 Web Page to add voter.

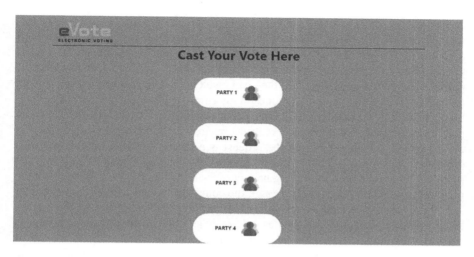

FIGURE 9.5   Voting Web Page.

Once all the procedures are followed properly, then on the day of election, the authority can allow users to vote and after a certain period of time it can end voting. Figure 9.5 shows the voting page. The result of elections can be found as soon as the voting ends by just clicking the button. This contract tries to make elections more secure and an extremely fast process.

9.7   THE USE OF IPFS

In eVote, IPFS plays a very significant role. First, a user registers to get validated to cast his/her vote. An eligible voter is required to upload an identity document that is recognized by the government. This document is shared using the IPFS technology.

When the eligible voter uploads the document on IPFS, a hash id is received, which corresponds to the address of the document. A function ("ad dVoter") takes the hash id generated by the IPFS upon sharing/uploading a file. On a successful function call, a request to both the controlling officials is generated. Controlling authorities have the hash id of the corresponding eligible voter and check his/her details against the information stored in the government system and takes the required action, whether to approve or reject the request. If the request is accepted by both the controlling officials, in that case, the corresponding voter is allowed to cast his/her vote, using his unique hash id.

9.8   THE INTERFACE

The interface we have created is a web interface that uses ReactJS as the framework. The web site will be all-screen compatible and will run efficiently even on slower devices. The interface consists of three parts: one for the voters and parties, the other for the authorities, and the third part for the elections. Initially, there are two input fields in which authorities have to insert their hash id and purpose it for creating and deploying the contract. Once the contract is deployed, we get a link for three different phases as mentioned above. The first page will allow parties to append their agenda. If there is a new party that wants to enter elections, then it can also request the authority to add their names to the parties' list. The voters can also check all the parties and see their agendas, and they can also check if their name is on the voter list or not. New voters can also request authorities to add their names to the voter list by providing the required hash code generated by IPFS for their documents. The authorities on their end receive all requests regarding adding new voters or parties. Authorities will then verify that the request is a genuine request or not, and then only, the voter will be added to the voter list or the new party will be added to the party list. Once the time for adding new parties and voters finishes, we move to the second phase, where the authorities sign the commencement of elections using their signatures (hash ids), and then the voters will be allowed to cast votes for the parties. After a certain period, the authorities will again sign the contract to stop voting. As soon as the voting is over, authorities can call the function of results, and the results are displayed immediately on the website, where everyone can see the results. It can make the online conduct of elections more secure and quick (Figure 9.6).

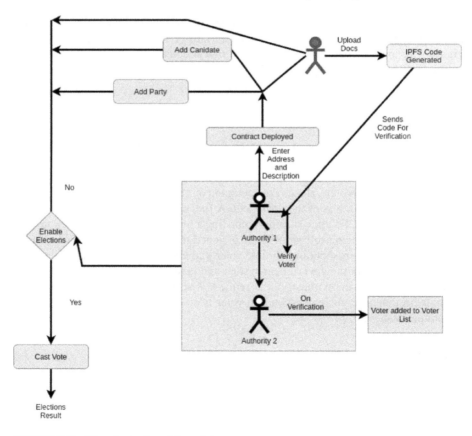

FIGURE 9.6 Flowchart of working of the system.

9.9 RESULTS

From the facts stated in the paper, we have deduced a table that compares traditional systems that are discussed in Section 9.2 and the proposed system in Section 9.5 (Figure 9.7).

To understand the view of people on current election procedures and to understand their thoughts on online voting, we surveyed seven, different states of India that is, Assam, Delhi, Jammu, Madhya Pradesh, Odisha, Rajasthan, and Uttar Pradesh. We received a large number of responses, and the feedback received is shown below. In our survey, we have considered only those who have access to the internet because they are aware of the pros and cons of the internet, and so would provide us better responses and thoughts on the system we have created.

From Figure 9.8, it is observed, that 56.2% of the people who took part in the survey belong to the age-group of 18–30 because they are the people who have

Characteristics	Ballot Paper	Electronic Voting Machine(EVM)	eVote (Propose System)
Transparency	Low	Low	High
Security	Low	Moderate	High
Data Manipulation	High	Low	Not Possible
Counting Mistakes	High	Very Low	Not Possible
Vote Stuffing	High	Low	Not possible
Counting Time	High	Low	Instantaneous
Accountability	Low	Low	High
Vote from home	Not Allowed	Not Allowed	Allowed

FIGURE 9.7   Comparison of ballot paper, EVM and e-vote.

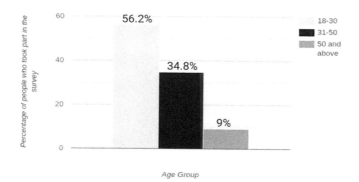

FIGURE 9.8   Age group of people who took part in the survey.

more access to internet, thus they have better knowledge about working of the internet. These people are also the next generation of voters, so their suggestions and responses are paramount.

From Figure 9.9, it is concluded that the survey is not gender biased.

Figure 9.10 shows that 84% of the users have daily access to the internet. The internet is a necessity these days and has a very high potential in shaping the world, thus, we should tap into that potential. Thus, we took the initiative to bring elections online.

Figure 9.11 shows that over 82% of the people who took part in the survey are regular voters. Thus, we can infer that they are aware of the elections and its procedure and also know about the regular procedures that are being followed in the elections.

Figure 9.12 shows that around 82% of the people have voted in the last elections, and since they have voted recently, they will provide the data required for our research. At the same time, they are more aware of the current situation. We then asked these people a few questions regarding internet and how has it helped them throughout the recent elections.

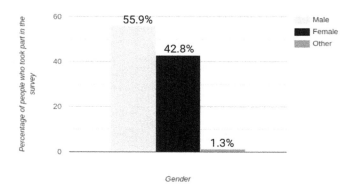

FIGURE 9.9   Gender ratio of people who took part in the survey.

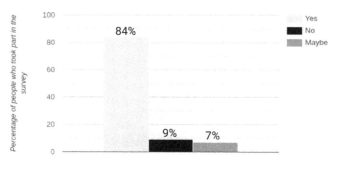

FIGURE 9.10   Number of people who are regular internet user.

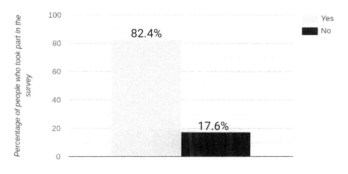

FIGURE 9.11   Number of people who vote regularly in elections.

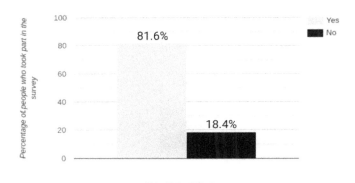

FIGURE 9.12   Number of people voted in the last elections.

Figure 9.13 shows that around 91% of the people believe that the internet, in some way, has made them aware of the current scenarios of elections.

Figure 9.14 shows the response of people to the questions mentioned in the figure. More than 60% of people accept the fact that internet has helped them in taking a political decision that has helped them in choosing the candidates they have voted for in the elections.

We asked people whether they are satisfied with the amount of money that is spent to conduct the elections. This question was asked because in this paper, we are proposing a system that is going to reduce the expenditure incurred on elections by a large margin and thus, we want to know what are the thoughts of the voter on this topic. Figure 9.15 shows that over 80% people are not satisfied with the money that is being spent, and this ensures that online voting can overcome this part and can increase voters' satisfaction.

An instantaneous system is proposed in this paper that takes only a few seconds for a voter to cast a vote. The voter has to do a press some buttons on his device to

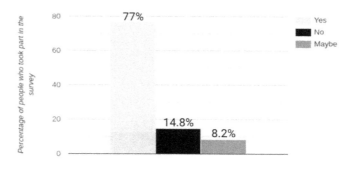

FIGURE 9.13 Ques: Has internet made you aware of political scenarios?

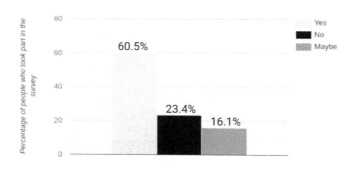

FIGURE 9.14 Ques: Has internet helped you to choose which candidate to vote?

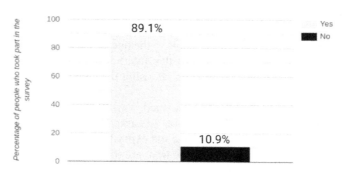

FIGURE 9.15 Ques: Are you satisfied with a large amount of money spent to conduct elections?

cast his vote. However, in the EVMs or ballot paper system, people have to stand in queues for their turn. From Figure 9.16, it is observed that over 71% of the voters are not satisfied with the time taken to cast a vote through EVMs and ballot paper system.

Many arguments raise doubts on the transparency in traditional voting systems. From Figure 9.17, it can be inferred that more than 63% of people do not find traditional voting systems to be transparent.

To know whether people think that there should be an option to vote online we asked this question to the people and from Figure 9.18, we can say that over 81% of people are in favor of having an option of online elections.

The more the number, the better the results. It is believed that more voters ensure the quality of elections, and in Section 9.2, we have seen into the issues faced by people, which result in decreasing the vote count. Since people these days are on the internet, we asked people that if online voting is introduced, will it increase the

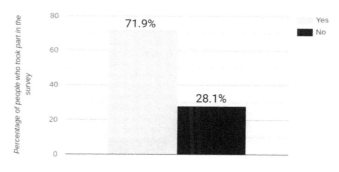

FIGURE 9.16 Ques: Do you think that it is time-consuming to stand in queues at polling booths?

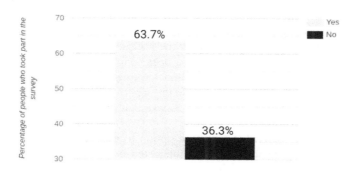

FIGURE 9.17 Is traditional voting system transparent?

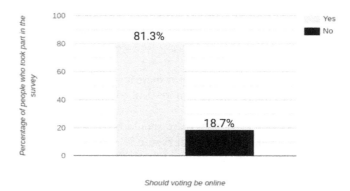

FIGURE 9.18 Ques: Do you think there should be an option of being able to vote online in elections?

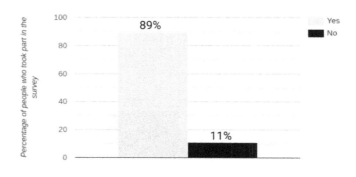

FIGURE 9.19 Ques: Can online voting encourage more voters to vote?

number of voters, and Figure 9.19 shows that around 90% of people believe that online voting will increase vote count. This shows that the proposed system will not only satisfy people on different parameters, but will also increase the number of voters casting their votes. To get to know the mindset of people about the security of online voting system, since there have been many incidents in the past where allegations have been made that the traditional voting system was not secure, so we asked them what are their views regarding the security of online voting system and Figure 9.20 shows that over 71% people believe that online voting can be a secure medium.

Figure 9.21 shows that when given an option to choose between online voting and traditional voting, people are more interested in voting online. In the system proposed in the paper, we have also created a system using IPFS where people can securely put their documents that can be verified easily by the concerned authorities. We asked people that if such a system is provided, will they use that system. Figure 9.22 shows that around 46% of people are willing to use the system, whereas

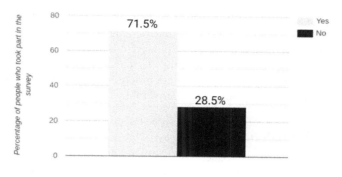

FIGURE 9.20 Ques: Do you think online voting can be secure?

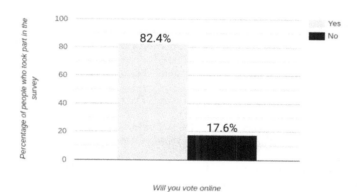

FIGURE 9.21 Ques: If there is also an option of online voting in the next elections, then which mode will you choose?

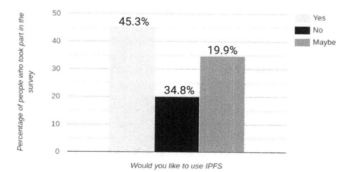

FIGURE 9.22 Ques: Would you like to save your voter id data on a platform (IPFS) where only you can access it and you can give permission to others to access it?

34% are unsure for now. After this survey, we sent an invite link to the participants on their email id for the dummy election that we conducted, using the system proposed in the paper. We received a good response from them. We asked them to rate their experience on a scale of 1–5 (1 being the worst and 5 being the best) we received an average of 4.1. This shows that the system proposed in the paper can be a better approach for conducting elections as compared to the traditional systems.

9.10   CONCLUSION

The system proposed in the paper combines two state-of-the-art technologies, Blockchain and IPFS, to overcome the flaws of traditional techniques that are in practice and also resolve the issues of transparency and security. The users of the system are themselves accountable for their actions. Thus, there is no scope that someone can manipulate data, and through this, we can conclude that the proposed system can save the literal values of the "Right to Vote" of people.

REFERENCES

1. G. A. O'donell, "Delegative democracy," *Journal of Democracy*, vol. 5, no. 1, pp. 55–69, 1994.
2. V. O. Key Jr, "A theory of critical elections," *The Journal of Politics*, vol. 17, no. 1, pp. 3–18, 1955.
3. E. Posada-Carbo, *Elections before Democracy: The History of Elections in Europe and Latin America*. Springer, 2016. New york
4. N. Chazan, "African voters at the polls: A re-examination of the role of elections in African politics," *Journal of Common-Wealth & Comparative Politics*, vol. 17, no. 2, pp. 136–158, 1979.
5. A. Blais, L. Massicotte, and A. Yoshinaka, "Deciding who has the right to vote: A comparative analysis of election laws," *Electoral Studies*, vol. 20, no. 1, pp. 41–62, 2001.
6. R. Barnes, "Vote dilution, discriminatory results, and proportional representation: What is the appropriate remedy for a violation of section 2 of the voting rights act," *UCLA Law Review*, vol. 32, p. 1203, 1984.
7. R. Johns and M. Shephard, "Facing the voters: The potential impact of ballot paper photographs in British elections," *Political Studies*, vol. 59, no. 3, pp. 636–658, 2011.
8. J. Kelley and I. McAllister, "Ballot paper cues and the vote in Australia and Britain: Alphabetic voting, sex, and title," *Public Opinion Quarterly*, vol. 48, no. 2, pp. 452–466, 1984.
9. P. L. Allen, "Ballot laws and their workings," *Political Science Quarterly*, vol. 21, no. 1, pp. 38–58, 1906.
10. G. Davies and J. E. Zelizer, *America at the Ballot Box: Elections and Political History*. University of Pennsylvania Press, 2015. Philadelphia, Pennsylvania
11. F. C. Schaffer, "Disciplinary reactions: Alienation and the reform of vote buying in the philippines," in *ponencia presentada en la conferencia "Trading Political Rights: The Comparative Politics of Vote Buying"*. Center for International Studies–MIT, Cambridge, 2002, pp. 26–27.
12. S. Nichter, "Vote buying or turnout buying? Machine politics and the secret ballot," *American Political Science Review*, vol. 102, no. 1, pp. 19–31, 2008.
13. J. N. Wand, K. W. Shotts, J. S. Sekhon, W. R. Mebane, M. C. Herron, and H. E. Brady, "The butterfly did it: The aberrant vote for Buchanan in palm beach county, Florida," *American Political Science Review*, vol. 95, no. 4, pp. 793–810, 2001.

14. B. B. Bederson, B. Lee, R. M. Sherman, P. S. Herrnson, and R. G. Niemi, "Electronic voting system usability issues," in *Proceedings of the SIGCHI Conference on Human Factors in Computing Systems*, ACM, 2003, pp. 145–152.
15. F. E. Lehoucq and I. Molina, *Stuffing the Ballot Box: Fraud, Electoral Reform, and Democratization in Costa Rica.* Cambridge University Press, 2002. New York
16. A. Ware, "Anti-partism and party control of political reform in the united states: The case of the Australian ballot," *British Journal of Political Science*, vol. 30, no. 1, pp. 1–29, 2000.
17. R. Rose and E. Tikhomirov, "Russia's forced-choice presidential election," *Post-Soviet Affairs*, vol. 12, no. 4, pp. 351–379, 1996.
18. N. G. Jayal, "Democracy in India," OUP Catalogue, 2007.
19. T. Kohno, A. Stubblefield, A. D. Rubin, and D. S. Wallach, "Analysis of an electronic voting system," in *IEEE Symposium on Security and Privacy, 2004. Proceedings. 2004.* IEEE, 2004, pp. 27–40.
20. R. J. Boram, "Electronic voting machine and system," Feb. 3 1987, US Patent 4,641,240.
21. W. Drechsler and U. Madise, "Electronic voting in Estonia," in *Electronic Voting and Democracy*, Springer, 2004, pp. 97–108.
22. D. A. Gritzalis, *Secure Electronic Voting.* Springer Science & Business Media, 2012, vol. 7. New york
23. J. Bannet, D. W. Price, A. Rudys, J. Singer, and D. S. Wallach, "Hack-a-vote: Security issues with electronic voting systems," *IEEE Security & Privacy*, vol. 2, no. 1, pp. 32–37, 2004.
24. C. Zucco Jr and J. M. Nicolau, "Trading old errors for new errors? The impact of electronic voting technology on party label votes in brazil," *Electoral Studies*, vol. 43, pp. 10–20, 2016.
25. J. M. Davis III and S. Thomas, "Direct recording electronic voting machine and voting process," Dec. 10 1996, US Patent 5,583,329.
26. E. A. Fischer, K. J. Coleman, S. Resources, I. Division, Government, and F. Division, "The direct recording electronic voting machine (dre) controversy: FAQS and misperceptions," Congressional Research Service, Library of Congress, 2007.
27. M. F. Mursi, G. M. Assassa, A. A. Abdelhafez, and K. M. Abosamra, "A secure and auditable cryptographic-based e- voting scheme," in *2015 Second International Conference on Mathematics and Computers in Sciences and in Industry (MCSI).* IEEE, 2015, pp. 253–262.
28. D. Basin, H. Gersbach, A. Mamageishvili, L. Schmid, and O. Tejada, "Election security and economics: It's all about eve," in *International Joint Conference on Electronic Voting.* Springer, 2017, pp. 1–20.
29. B. Ondrisek, "E-voting system security optimization," in *2009 42nd Hawaii International Conference on System Sciences.* IEEE, 2009, pp. 1–8.
30. D. A. Gritzalis, "Principles and requirements for a secure e-voting system," *Computers & Security*, vol. 21, no. 6, pp. 539–556, 2002.
31. M. Swan, *Blockchain: Blueprint for a New Economy.* O'Reilly Media, Inc., 2015. Sebastopol, City in California
32. F. Hjalmarsson, G. K. Hreiarsson, M. Hamdaqa, and G. Hjalmtysson, "Blockchain-based e-voting system," in *2018 IEEE 11th International Conference on Cloud Computing (CLOUD).* IEEE, 2018, pp. 983–986.
33. L. W. Cong and Z. He, "Blockchain disruption and smart contracts," *The Review of Financial Studies*, vol. 32, no. 5, pp. 1754–1797, 2019.
34. G. Wood, "Ethereum: A secure decentralised generalised transaction ledger," *Ethereum Project Yellow Paper*, vol. 151, no. 2014, pp. 1–32, 2014.
35. C. Dannen, *Introducing Ethereum and Solidity.* Springer, 2017, vol. 1. New York
36. J. Benet, "IPFS-content addressed, versioned, p2p file system," arXiv preprint arXiv: 1407.3561, 2014.

37. M. S. Pillai, G. Chaudhary, M. Khari, and R. González Crespo, *Real-Time Automatic Automobile Accident Detection through CCTV Using Deep Learning, Soft Computing.* Springer, 2020

38. V. Srivastava, S. Srivastava, G. Chaudhary, and F. Al-Turjman, "A systematic approach for COVID-19 predictions and parameter estimation," *Personal and Ubiquitous Computing*, 1–13, 2020. https://doi.org/10.1007/s00779-020-01462-8

39. H. M. R Afzal, S. Luo, M. K. Afzal, G. Chaudhary, M. Khari, and S. A. P. Kumar, "3D Face reconstruction from single 2D image using distinctive features," *IEEE Access*, vol. 8, pp. 180681–180689, 2020.

10 Nessus
A Vulnerability Scanner Tool in Network Forensic

Himanshu
Ambedkar Institute of Advanced Communication
Technologies and Research

CONTENTS

10.1 INTRODUCTION

To understand the network vulnerability attack, one should understand that the attacker doesn't attack in isolation, rather than that, they attack in a combination. So, this chapter presents the study about the history of Nessus, what is it and how it works with the plugins. It also studies about the features of Nessus which makes it a highly recommended network vulnerability scanning tool in the network forensic. This chapter also presents how Nessus is to be downloaded and the steps involved. A block diagram also describes the flow of Nessus vulnerability scanner. Nessus is helpful in digital twinning because of its feature of network vulnerability risk assessment and it is also reduces the assessment time, which may cause an increase in updating the security protocols in digital twin evaluation. By using the feature of Nessus, the simulation of the real-world data to check network vulnerability scanning becomes easy.

10.2 HISTORY OF NESSUS

Renaud Deraison introduced Nessus project in 1998 when he was only 17 years old for the internet community to provide free remote security scanning. He introduced Nessus as an open-source project, led by the community while he was pursuing his career in IT field. So, the copyright of Nessus belongs to the Renaud Deraison. The availability of the source code to all has led to the creation of forks, which are the rivals to the Nessus system. Soon it became the leading vulnerability scanner in the world. Tenable network security company co-founder Renaud Deraison changed the Nessus 3 to a licensed version. The minority of the plugins and the Nessus 2 version are still GPL, which leads to the open-source project, based on Nessus like porz-wahn and openvas. Tenable began working in 2002. It had 2 million downloads of the free version at that time and 27000 businesses were already using it, while the paid version of Nessus came in 2005.

10.3 RELATED STUDY

Nessus is a network vulnerability scanner [1,2]. It uses plug-ins, which are generally separate files, and the vulnerability checks are handled by them. Plug-ins are the individual pieces of codes which are to be used by Nessus for conducting individual scan on target plug-ins that are wide in their capabilities and number. At the same time, digital twin is a computer program operation, which helps us in creating simulation of real-world data and predicts the network software performance or other software performances, along with internet of things analytical software and artificial intelligence to enhance the performance.

10.4 BASICS

It basically makes a virtual computer that receives feedback from the servers. Its configuration is simple or complex, according to the requirements. So, the plug-ins are launched in the manner given below and target the host. Figure 10.1 shows the scan functions in which, firstly, the parameters of scan are defined and then click on the new scan to create a new scan. After that, add the necessary details and then launch the scan. A report gets generated which shows that how many vulnerabilities are present in the network.

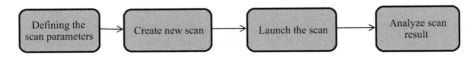

FIGURE 10.1   Scan functions of Nessus.

1. Firstly, identify which OS and services are running on which port.
2. Then, identify which of the software component is vulnerable to attack.
3. Identify if the requirements are met on various hosts.

Nessus generally uses server-client architecture [3,4], and the main server of that will need to be built on a supported UNIX operating system. The client is always available for the windows, UNIX, and Linux. The server is not taken as an option because it performs security checks. The user account set up is done by the administrator and for other team members and it also issues rights to those accounts. The scans are only done after the client logs on to the server. Nessus is used by students and organizations for checking the vulnerabilities in the network.

If a person wanted to hack your local network, then the first thing they have to do is run a vulnerability scan, and then they have to run the penetration test. A vulnerability scan looks and digs through several devices on your network and looks for potential holes, like outdated software with known vulnerabilities, open ports and default passwords on devices and the saved passwords on the device. If they find anything from these, then a hacker would surely test those vulnerabilities, and then he will find a way to exploit them. Testing those vulnerabilities is a two-step process because a scan only reveals the chances of the problems and the penetration test verifies if the problem is actually exploitable or not. The upcoming Sections 10.4.1–10.4.3 describe the Nessus versions and Section 10.5 describes the features of Nessus. Section 10.6 describes block diagram of Nessus vulnerability scan, and Section 10.7 describes how Nessus is to be downloaded and how it works in steps. Section 10.8.1 describes the pros of Nessus, and Section 10.8.2 describes its cons.

As the Nessus is a commercial software which is made for scanning the vulnerabilities, its free version offers plenty of tools to help and explore your home network. Nessus is available in three versions:

1. Nessus essential
2. Nessus professional
3. Tenable.io

10.4.1 Nessus Essential

It is the free version of the vulnerability scanner. Its scans are limited to 16 IP addresses and this tool is basically aimed for students of the network technology and information security. It is also used by the businesses for understanding the requirements and see how it works for their purpose and then the organization buys the paid version. So, it is not only for the students but it is also used by the organizations and home users. You can pick up the free plug-ins that are provided by the community on the internet.

10.4.2 Nessus Professional

It is one of the paid versions of the vulnerability scanner. It gives you full support features. It is the same software as Nessus essential but here there is no restriction limit of 16 IP addresses. It gives you live results and the system also sweeps periodically. Nessus professional [5] is charged by the subscription method. This means that it is an yearly service and there is no monthly payment method. If you want discount, then you have an option for multiyear subscription which is generally 3 years long. It provides seven days' free trial version.

10.4.3 TENABLE.IO

It is a cloud version of Nessus pro. It is more costly than Nessus professional and it comes with advance support packages. It starts with the base price of 65 nodes and the price is increased in accordance with the node increment.

10.5 FEATURES OF NESSUS

1. One of the most attractive and effective features of Nessus is that it is freely available. Anyone may download it and then use Nessus essential software. But there are some paid versions also available.
2. It is an open-source software and many people contribute to Nessus every day that helps it to stay up-to-date and freely available at all the times for the user.
3. Plug-ins can be updated once the features of plug-ins with new vulnerabilities are released.
4. Nessus scans for vulnerabilities on Linux, Windows and UNIX systems. This will help Nessus to become a good, all- round tool so that you can scan in a mixed environment in single session.
5. Nessus also utilizes N-map for port scanning.
6. The server-client architecture is a plus point of Nessus if there are multiple persons using the system. So, the user rights can be defined to lock down the types of scans they can do and unlock the ones they don't want to do.
7. The plug-in architecture of Nessus ensures that it checks each vulnerability as an individual plug-in. This means that it gives you the right to your own plug-ins.
8. It doesn't do the penetration testing because its main focus is to scan the system or the applications.
9. Nessus allows multiple profiles/policies to do different types of scans such as malware scan, bad shell shock detection and web applications scan.
10. It classifies the vulnerabilities into a prioritized risk-based-categories, from critical to normal.

10.6 BLOCK DIAGRAMS OF NESSUS SCAN PROCESS

The block diagram of Nessus shows how the user interacts with Nessus software and performs the vulnerability scanning onto that. Figure 10.2 shows how Nessus scan works and describes how the Nessus scan process works in which the user can check

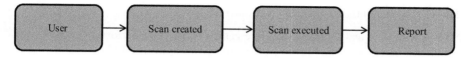

FIGURE 10.2   Block diagram of Nessus scan.

the vulnerability of network IP address locally and remotely. With the help of the above figure, the user can select which type of scan he wants to do and then add the necessary details to the new scan. Once the scan is executed, a report is generated, which illustrates the vulnerabilities present in the network.

10.7   PROCESS OF NESSUS

10.7.1   STEP ONE: DOWNLOAD AND INSTALL NESSUS

For downloading Nessus, the users first need to sign up for an online account and then they can download the software and get an activation code on the email-id filled at the time of registration. The steps involved are:

1. First, go to Nessus homepage, then enter the name and email address, and then click on the Register button. Users have to enter a real email address here because Nessus sends an activation code on that email id to be entered in the registration form.
2. Now, click the Download button, to download Nessus on your operating system, according to the system bits. It's available for Mac, UNIX, windows and Linux.
3. When the downloading is complete, then run the installer package and follow instructions on the screen and finish the installation.
4. Now, Nessus creates a local server on your computer or laptop and runs from there.

10.7.2   STEP TWO: SET UP YOUR NESSUS ACCOUNT AND ACTIVATION CODE

When Nessus is installed on your computer then point the web browser to the following address: https://localhost:8834/. This is the address where users have to complete the signup process and activate his/her copy of Nessus.

1. When the user launches Nessus for the very first time, then he/she gets a warning like "Your connection is not secure" from your browser. So, click "Advanced" option and then select "Proceed to local-host" option to bypass this warning.
2. Now, Nessus will download a number of plugins and tools so that it can correctly scan the network with updated utilities available. This downloading process along with compilation of plug-ins [7] take some time.

10.7.3   STEP THREE: START A VULNERABILITY SCAN

Now, it is time to test the network. Here generally, the user performs basic network scan because it gives a complete overview of network vulnerabilities.

- **A. Click on "New Scan"**: Here, the user clicks on the new scan after logging into his/her account. Then, the dashboard of new scan appears as shown in Figure 10.3
- **B. Click on "Basic Network Scan"**: After clicking on new scan, the user has to select which type of scan needs to be performed, as shown in Figure 10.4. Generally, the basic network scanning is done in the Nessus.
- **C. Name the scan and add a description about it and also set targeted IP addresses**: After selecting Basic network scan, the user has to fill the necessary details like name of the scan, along with its description and also set the target. This is shown in Figure 10.5.
- **D. In "Targets" field, the user has to enter the IP scanning details about home network or local network.** For example, if our router is at 192.168.0.1, then after setting target and launching the scan, a report of network vulnerabilities is generated in which according to the color code, the effect of vulnerability is accessed. This is shown in Figures 10.6 and 10.7

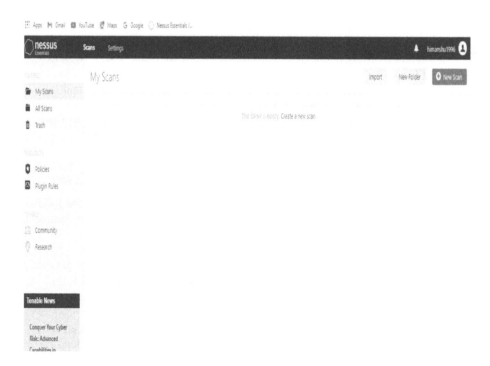

FIGURE 10.3   New scan dashboard.

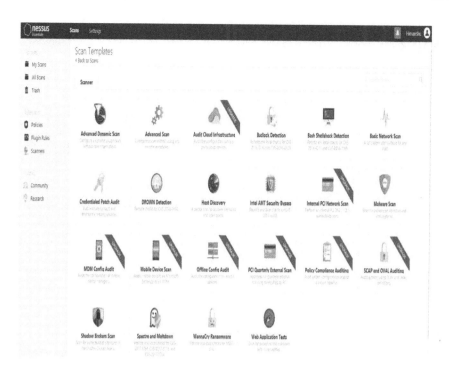

FIGURE 10.4 Basic network scan.

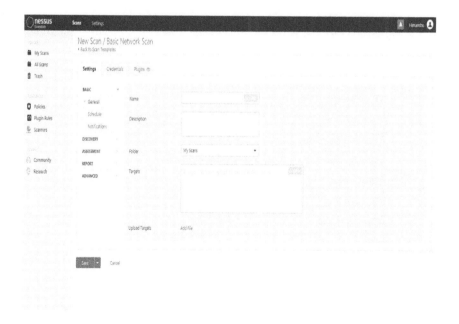

FIGURE 10.5 Add the necessary details.

FIGURE 10.6   Local network vulnerabilities.

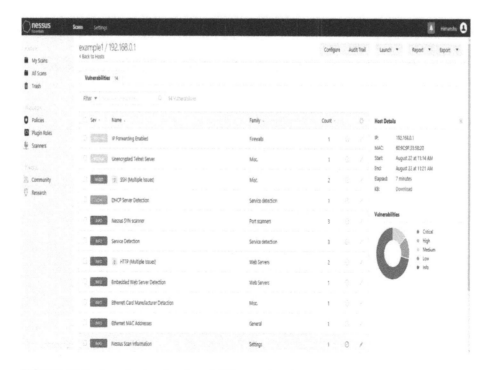

FIGURE 10.7   Local network vulnerabilities continued.

E. **In "Targets" field, the user has to enter the IP scanning details about remote network.** For example, if our router is at 192.168.1.52, then after setting the target field, a report gets generated, in which the vulnerabilities of the network are shown, which also depicts its effectiveness according to the color code. This is depicted in Figures 10.8–10.10.

F. **Click on "Save":** Now, depending on how many devices and what type of devices are connected to your network, it will take time.

There are more than one scanning methods in Nessus essential. Apart from the Basic Network Scan, you can also run an Advanced Scan that includes more parameters to filter your search, A Bad lock Detection scan hunts down a security issue with SAMBA. A Shellshock scan looks for the vulnerabilities in Mac machines and old Linux. A DROWN scan basically looks for the computers that are hosting sites which are susceptible for DROWN attacks. Apart from these, there are few more scans that are also available in Nessus essential, as shown in Figure 10.11

10.7.4 STEP FOUR: MAKE SENSE OF THE RESULTS

Figure 10.12 shows plugin features of Nessus in which there are certain set of rules created, according to requirements of the user. Figure 10.13 shows download logs options like basic and balanced which also contain their own feature set. According to the feed status download, logs are maintained.

FIGURE 10.8   Remote network.

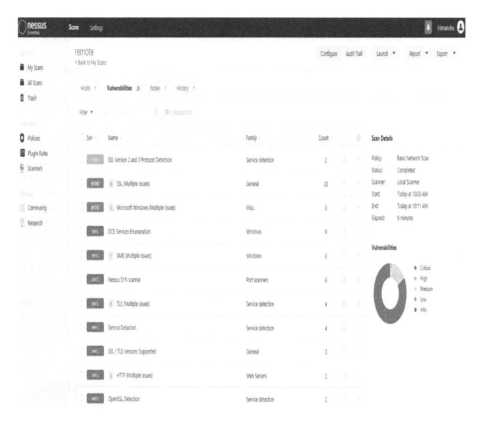

FIGURE 10.9   Remote network vulnerabilities.

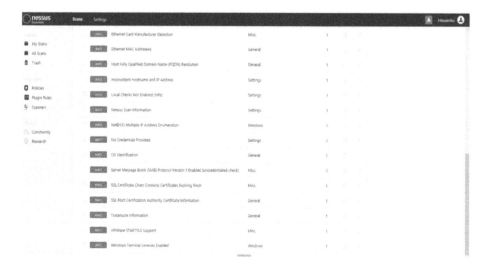

FIGURE 10.10   Remote network vulnerabilities continued.

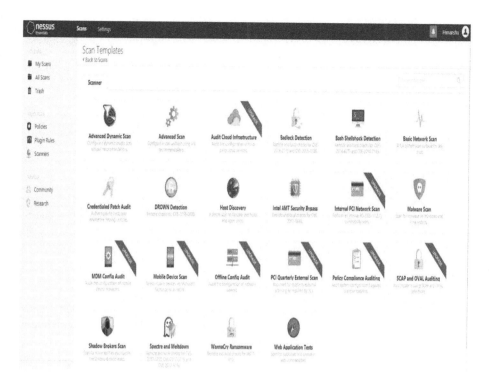

FIGURE 10.11   More features of Nessus.

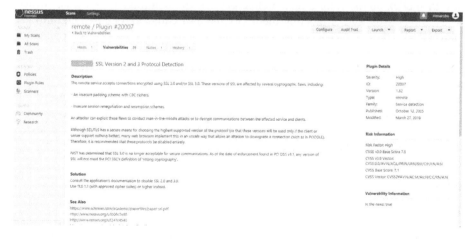

FIGURE 10.12   Plugins.

FIGURE 10.13 Download logs.

Once Nessus finishes off his task, then user will see a bunch of color-coded graphs for each of the devices of your network. Each color of the graph signifies a different level of danger of vulnerability, from critical to low.

1. Red represents Critical
2. Yellow represents Medium [7]
3. Green represents Low
4. Blue represents information

Results should include all the devices available on local network, from the router to the Wi-Fi enabled printer. If the user wants more information about the vulnerabilities, then click on the color graph. Vulnerabilities are identified as "plugins," which is just the Nessus way of labeling them. So, the user can click on any plugin to get more information about the vulnerability, including press releases, white papers or patch notes.

If the user wants to know the network vulnerabilities as a whole, then vulnerabilities tab needs to be clicked. [8].

So, there is a chance that some of these vulnerabilities will be a bit obvious, which means that vulnerabilities are always present. For example, Nessus picks up any one device, then it is still using the points or default password as the computer or device is running on an outdated firmware.

10.8 PROS AND CONS OF NESSUS

10.8.1 PROS OF NESSUS

1. Missing critical patches of a server or workstation can be found with the help of Nessus.

2. It does not only list the vulnerabilities but also describes the vulnerability in detail.
3. It also does the PCI scanning in Nessus professional.
4. Nessus also has the feature of multiple profiles/policies to do different types of scan such as malware scan, bad shell shock detection and web applications scan.
5. It also has the feature of classifying the vulnerabilities into prioritized risk-based categories, from critical to normal.
6. The plug-in architecture of Nessus ensures that it checks each vulnerability as an individual plug-in. This means that it gives you the right to your own plug-ins.

10.8.2 Cons of Nessus

1. It could use the upgrade within the records.
2. Sometimes, scans can take a very long time to complete. So, we have to break them down in small sections.
3. The disabling of plugins inside the plugins group is not allowed for advance users.
4. Ongoing status of scan may improve because it either shows 0% or 100%, which means that how much percentage of scan is complete is not shown by the progression bar.
5. Scanning can be simplified further, with the help of default setting configuration.

10.9   CONCLUSION

This chapter presents study of the history of Nessus, what it is and how does it work with the plugins. It also studies features of Nessus which makes Nessus a highly recommended network vulnerability scanning tool in the network forensic. A block diagram also describes the flow of Nessus vulnerability scanner. This chapter also presents that the Nessus helps in digital twin evaluation for the network vulnerability scanning purpose, to make the simulation of the real-world data, to predict the network vulnerability assessment performance.

This chapter also presents how Nessus is to be downloaded and the steps involved in taking a local IP address (192.168.0.1) and checking for the vulnerability. We found 2 vulnerabilities in medium state, 2 vulnerabilities in low and the rest 16 lie under info category. Then we performed the same with a remote IP address (192.168.1.52) and we found 2 vulnerabilities in high, 10 in medium state and the rest 72 lie under info category.

REFERENCES

1. A. Sowmyashree and H. S. Guruprasad. "Evaluation and analysis of vulnerability scanners: Nessus and OpenVAS." *International Research Journal of Engineering and Technology (IRJET)*, Bangalore, India, Vol. 7, No. 5, pp. 2068–2073, 2020.

2. Sandeep Kumar Yadav, Daya Shankar Pandey and Shrikant Lade, "A comparative analysis of detecting vulnerability in network systems." *IJARCSSE*, Vol. 7, No. 5, pp. 336–340, May 2017.

3. Sheetal Bairwa, Bhawna Mewara and Jyoti Gajrani. "Vulnerability scanners: A proactive approach to assess web application security." *International Journal on Computational Sciences & Applications (IJCSA)*, Vol. 4, No. 1, pp. 113–124, Ajmer, India, February 2014.

4. Peng Li and Baojiang Cui, "A comparative study on software vulnerability static analysis techniques and tools." In *Proceedings of the IEEE International Conference on Information Theory and Information Security (ICITIS)*, pp. 521–524, 2010.

5. Kushe R. "Comparative study of vulnerability scanning tools: Nessus vs Retina." *International Scientific Journal "Security & Future"*, Albania, Vol. 1, No. 2, pp. 69–71, 2017.

6. Harry Anderson. *Introduction to Nessus*. INFOCUS, 2003. http://apachepersonal.miun.se/~janjon/oldcourse/dtab80/lab/lab2/nessus_1.pdf

7. Harrison, Lane, Riley Spahn, Mike Iannacone, Evan Downing, and John R. Goodall. "Nv: Nessus Vulnerability Visualization for the Web." In *Proceedings of the ninth international symposium on visualization for cyber security*, pp. 25–32. 2012.

8. Paul Schmelzel. *Nessus: Vulnerability Scanning and beyond*. SANS, United States, 2002.

11 Case Studies Related to Depression Detection Using Deep Learning Techniques

Jagannath Jayanti, Madhavendra Singh, and Vibhuti Jain
Guru Gobind Singh Indraprastha University

CONTENTS

DOI: 10.1201/9781003132868-11

11.1   INTRODUCTION: BACKGROUND

In the last decade itself, mental health issues have increased manifold. Amongst these, depression is more predominant. Depression has been a culprit in deterring happiness, relationships, careers or in a nutshell, the lives of not only the individuals affected with them but also of their families and close ones. People battling depression unknowingly start to isolate themselves. They gradually stop indulging in activities they once used to enjoy. They begin to go into solitary, backing away from their social lives. Depression among youth of the society has significantly increased more than other segments of the society. Depression can be caused due to numerous reasons like abuse, death or loss of loved ones, new changes in life, like starting a new job, moving to a new place, joining a new school or college in case of students, substance abuse like consumption of drugs, or because of a feeling of alienation from closed ones. Depression can also exist along with some other major illnesses.

Depression can be fatal because of the direct implications it has on an individual, and so it needs to be prevented. Since, prevention of depression is very difficult, due to numerous reasons causing it, methods to cure it have to be developed. In case of depression, early detection can aid in curing it relatively easily, compared to later stages. It can have huge impacts on the mindset of the suffering individual. Negativity can be replaced with healthy thoughts, positive emotions etc.

In the case studies shown later in this chapter, the proposed solution is inspired by the fact that people suffering with depression isolate themselves and are not able to ask for help from family, friends or doctors. Also in many cases, the person might not know that he/she is slipping into depression. Hence, a system is proposed wherein an individual can upload an image of him/herself or give a text input and by using a convolutional neural network or/and a recurrent neural network respectively based on the input, a prediction is made as to whether that person may or may not be showing any signs of depression.

Since Convolutional and Recurrent Neural Networks are used in the case studies for making predictions on images and text respectively, let's look at CNNs [1] and RNNs [2] before diving into the case studies.

In the following session, first basic concepts related to convolutional neural networks are explained, followed by explanation of Long Short-Term Memory (LSTM), a type of recurrent neural network which is used in the case study. Considering the above sections, there are two case studies related to image processing and natural language processing, employed for depression detection.

11.2   RELATED CONCEPTS

11.2.1   CONVOLUTIONAL NEURAL NETWORKS (CNN)

Artificial Intelligence has witnessed an exponential growth in bridging the gap between capabilities of mankind and machines.. One such area is the field of Computer Vision. The aim of this field is to empower machines to perceive the world as humans do, and use this knowledge for heaps of tasks like image and video analysis, image recognition and classification, etc. The evolution in Computer Vision has been made with one algorithm – A Convolutional Neural Network.

A Convolutional Neural Network is a class of deep learning neural networks which takes in an image as input, assigns weights (importance) to various aspects of the image and distinguishes one image from another [3]. The structure of a CNN is symmetrical to that of the connectivity of neurons in human mind. An image is a sequence of pixels arranged in a matrix/grid like fashion that contains values to denote brightness and color of that pixel. For training and testing, each input image is passed through a set of convolution layers with filters, etc. A typical CNN comprises three layers: a convolutional layer, a pooling layer, and a fully connected layer [4]. Through these filters, CNN successfully preserves Spatial and Temporal dependencies of an image.

11.2.1.1   Convolution Layer

This is the core layer to extract features from input images such as vertical & horizontal edges, etc. It conserves the relation between pixels by learning image features (by performing dot product between two matrices, where there is a set of parameters, i.e., kernel, and the other is input). Kernel is spatially smaller than the input image but is denser. For instance, if an image consists of BGR channels, the kernel height and width will be spatially small, but the depth can extend. During execution, the kernel slides across the width and height of the input image, producing a two-dimensional representation, also known as activation map [5]. This layer can generate two types of results:

- Dimensionality of feature is reduced as compared to input (by applying Valid Padding)
- Dimensionality of feature remains the same or is increased (by applying Same Padding)

In trivial neural networks, every output unit interacts with each of the input unit. However, CNN has limited interaction (by making kernel smaller than input). For instance, an image can be made of millions of pixels, but on preprocessing it with kernel, we detect only meaningful information that consists of thousands or hundreds of pixels, helping us to reduce memory requirement of model, and at the same time, also improves efficiency of model [6].

- **Strides**: It can be considered as the jump a filter takes, while convolving with the image. It is the number of pixels a filter shifts across the image after performing the convolution operation with certain pixels.

11.2.1.2 Pooling Layer

This layer is somewhat similar to Convolution layer, but they only record precise position of feature and any small movement in position of feature will result in a different feature map. However, pooling provides translation in variance, i.e., an object would be recognizable no matter where it appears on the frame of image. Pooling layer runs on each feature map independently to create a new set of the same number of feature maps [7]. It gives a summarized version of the detected features in the input image and helps reduce overfitting and memory footprint. There are various pooling functions [8] such as:

- **Max pooling**: Function takes the maximum element from the feature map. This selects the brighter pixels from the image, that can be used when the background is dark and we want a lighter pixel.
- **Average pooling**: Function takes the average value from the feature map. It smoothens out the image, thus sharp features may not get identified.
- **Sum pooling**: Function takes the sum of all elements of the feature map. It measures the mean value of existence of a sample in a region.

The convolutional and pooling layer together form ith layer of a basic CNN structure. Depending on the input images, the number of layers may be increased for capturing minute details or decreased for capturing only overview of images, but it is possible only at the cost of computational power and memory. Pooling subsamples the images in its width and height but the number of channels remains the same.

11.2.1.3 Fully Connected Layer

These are feed forward neural networks; neurons have full connectivity with all other neurons in the preceding and succeeding layer. This layer is placed before the output of CNN and flattens our matrix into vectors. This feature vector holds information that is vital to the input. When the network gets trained, this feature vector is then used for regression, classification, or input to other networks [9]. In the final layer of fully connected layers, several activation functions can be used like SoftMax, reLu, etc.

11.2.1.4 A Few CNN Architectures

There are several architectures available which have played a crucial role in building algorithms that should power AI in the foreseeable future [10]. Some of them are:

- **LeNet**: It is a 7-level convolutional network that classifies digits. For processing high resolution images, more complex convolutional layers are required, so this technique is constrained by availability of computing resources.
- **AlexNet**: This module has a similar architecture as LeNet but is deeper/denser with stacked convolutional layers and more filters per layer.
- **ResNet**: Residual Networks uses Batch Normalization at its core. It makes use of identity connection, which helps to protect from vanishing gradient problems. This technique has the ability to train a neural network with up to 152 layers.
- **VGGNet**: This is a 16 convolutional layers network, which is currently one of the most preferred choices for extracting features from images.

- **GoogLeNet**: It is a 22-layer deep CNN. It is based on various small convolutions in order to considerably reduce the number of parameters. It is also used in batch normalization and image distortions.

11.2.2 RECURRENT NEURAL NETWORKS (RNN)

A simple neural network accepts a fixed size input which limits its usage in conditions which involve 'series' type inputs with no prior mentioned size. RNNs are engineered in a way that they can accept a series of inputs with no preset restraint on the size. One could ask what is the difference in the use case, and can I use a simple neural network again and again?

Of course one can, but the 'series' part of the input has to be given due emphasis here. A single input item from the series is correlated with other inputs and is very prone to influence its neighboring elements. Otherwise, it is just "many" inputs and not an input series. Hence, the need arises for something that can capture this relationship across inputs substantially.

Recurrent Neural Networks remember the past and their decisions are pretty much taken, considering what has been learnt in the past [11].

While RNNs learn in a similar fashion, like other neural networks, during training, they remember information learnt from previous input(s), while creating new output(s). It's part of the network. RNNs have the ability to take multiple inputs and generate multiple outputs and these output(s) are dictated not just by weights of inputs like a regular neural network, but also by a "hidden" state, consisting of the context based on previous input(s) or output(s). So, the same input could deliver a modified output based on what were the prior inputs.

11.2.2.1 Bidirectional RNNs

Sometimes, to predict the future, lessons learnt in the past do not work out, and one has to peek into the future to rectify the past. For example, in speech recognition, which may have ambiguity in just one section of the input data, the knowledge of what is succeeding is useful in order to gain an understanding of the data and help in correctly predicting the current instance [12].

This also introduces the question of "how further we need to look into the future?", the reason being, if we need to wait for a considerable amount of time to receive all the inputs in future, then the entire process becomes redundant.

11.2.2.2 Recursive Neural Networks

A Recurrent Neural Network (RNN) deduces the data inputs in a sequential order. Recursive Neural Networks are a more generalized form of plain RNNs. It operates on any hierarchical tree structure. Recurrent Neural Networks (RNNs) [13] also do the same, but their structure is more of a linear nature.

11.2.2.3 Encoder Decoder RNNs

Encoder Decoder RNNs are used in a wide variety of translation services. The idea behind them is two RNNs, of which, one is an encoder that has to update its states producing a single final "Context". This in turn is given to the decoder, which has to decipher this context to a series of outputs [14].

However, in this project, we have used LSTMs (Long Short-Term Memory) to classify text. So now, let's explore how an LSTM functions.

11.2.2.4 Long Short-Term Memory Cell (LSTM)

LSTMs are recurrent neural networks which are capable of learning order dependence to make correct predictions every time by using previous prediction values. LSTM is a special kind of recurrent neural network where understanding the context of data is very useful. Now the question is: what is a sequence prediction problem? Answers to this question can be understood through examples: for making language translations, for predicting next words in a smartphone keyboard, for recognizing stock market growths or crashes, etc. In today's world, these problems are the hardest to solve. With the creation of LSTM layers, these problems could be tackled easily [15].

LSTM networks are very much similar to RNNs with one major difference that unlike in the hidden layer updates in typical neural networks are replaced by memory cell updates. Because of this, LSTM is better at finding long-range dependencies in a dataset which is very important for sentence structures.

> **Input gate**: To update the current state of cell, the input gate is used. Previous state and current state are fed to a sigmoid function. Then it decides which values will be updated. 1 means important and 0 means not important. At last, the sigmoid output is multiplied with the "tanh" output.
>
> **Forget gate**: This gate decides as to which information should be kept to be passed on to the following data sequence. Data from the preceding state and data from the current input state is fed to the sigmoid function. Predicted values received are between 0 and 1. If the predicted value is closer to 0, it means to throw or reject the data, and if it is closer to 1, it means to keep/pass the data.
>
> **Output gate**: The output gate determines what the next hidden layer in the sequence will be. The hidden state layer contains the information of the previous inputs.
>
> **Cell state**: Now there is enough data to calculate the cell state. First, the cell state gets pointwise multiplied by the not so important vector (values of sigmoid output near to 0). This is done to drop some values in data that are not important.

This process is continued for all the data sequences in the dataset.

11.3 CASE STUDIES

Following are the two case studies wherein image processing and natural language processing techniques are used to detect depression.

11.3.1 CASE STUDY 1: DEPRESSION DETECTION USING IMAGE PROCESSING

For making predictions as to whether a person is depressed or not, we use CNNs. The following steps are taken in order to implement image processing [16].

11.3.1.1 Dataset

In this study, the "Facial expression detection image folders (fer2013)" dataset available on Kaggle is used. It can be found on the given link: https://www.kaggle.com/astraszab/facial-expression-dataset-image-folders-fer2013. This dataset consists of seven folders based on different emotions. These seven folders are then combined based on the emotions, resulting in two folders labelled 'depressed' and 'not depressed'.

11.3.1.2 Preprocessing

The large number of images collected (dataset) are not suitable to be given a neural network straightaway. It is due to this reason that they are said to be preprocessed.

For preprocessing, the following steps are taken:

1. The images are distributed into two folders labelled "depressed" and "not depressed", in order to achieve a labelled data to be fed into convolutional neural network. The image is converted into dimensions 32 * 32. Then all the images are converted into a matrix where each pixel is represented by a number. This is required because a machine can only understand numerical data.
2. Now the images are represented in image_data(X) format, that is the matrix form of all images and their corresponding labels (Y). It denotes "0" for "depressed" and "1" for "not depressed".
3. Since the images are RGB images, they have three channels, and they are divided by 255 to get values between 0 & 1. Since 255 (256 if 0 is not included) is the maximum value, dividing by 255 results in a 0–1 representation.
4. Then finally, the data is shuffled in order to be fed to the neural network.

Now that the images are preprocessed they can be fed to the neural network.

```
In [1]: import numpy as np
        import os
        from pathlib import Path
        from keras.preprocessing import image
        from keras.models import Sequential
        from keras.layers import *
        from keras.utils import to_categorical
        from keras.optimizers import SGD

In [46]: p = Path("images/data/train")

         dirs = p.glob("*")

         labels_dict = {"depressed":0, "not depressed":1}

         image_data = []
         labels = []

         for folder_dir in dirs:
             #print(folder_name)
             label = str(folder_dir).split("\\")[-1]
             print(label)

             for img_path in folder_dir.glob("*.png"):
         #        print(img_path)
                 img = image.load_img(img_path,target_size=(32,32))
                 img_array = image.img_to_array(img)
                 image_data.append(img_array)
                 labels.append(labels_dict[label])

         depressed
         not depressed

In [47]: image_data = np.array(image_data,dtype='float32')
         labels = np.array(labels)

         print(image_data.shape,labels.shape)

         (28709, 32, 32, 3) (28709,)
```

FIGURE 11.1 Preprocessing steps 1 and 2.

```
In [48]:  def preprocess_data(X):
              X = X.reshape((-3,32,32,3))
              X = X/255.0
              return X
          image_data = preprocess_data(image_data)
          print(image_data.shape)

          (28709, 32, 32, 3)

In [49]:  labels = to_categorical(labels)
          print(labels.shape)

          (28709, 2)

In [51]:  import random

          combined = list(zip(image_data,labels))
          random.shuffle(combined)

          #Unzip
          image_data[:],labels[:] = zip(*combined)
```

FIGURE 11.2 Preprocessing steps 3 and 4.

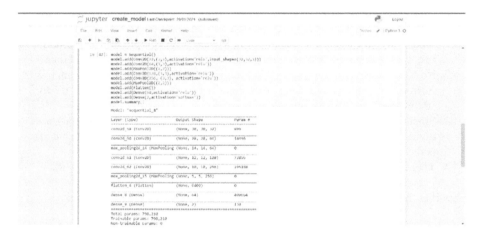

FIGURE 11.3 CNN architecture and summary for this project.

11.3.1.3 Convolutional Neural Network for Image Processing

1. The sequential model API of the Keras library is used to make a CNN [17].
2. Four convolutional 2D layers are made with 32, 64, 128, 256 filters respectively. Input size of each image is changed dimensionally to [32,32,3] after preprocessing of images. First, Conv 2D layer consists of 32 filters, with the dimension of each filter being [3,3]. Output of this layer is of dimensions [30,30,32] (here the third dimension is due to 32 filters which add 32 channels to the image). This is used as an input for the second Conv 2-D layer which contains 64 filters with the dimension of each filter being [3,3]. Output of this layer is of dimensions [28,28,64]. Now a MaxPool layer is added which helps to avoid overfitting of the model by reducing dimension of the output in the hidden layer. In this case, a Maxpool 2-D layer with dimension [2,2] is used. Output of this layer is of dimensions [14,14,64]. This is used as the input for the next Conv 2-D layer, which contains 128

filters with the dimension of each filter as [3,3]. Output of this layer is of dimensions [12,12,128]. This is used as the input for the next Conv 2-D layer which contains 256 filters with the dimension of each filter being [3,3]. Output of this layer is of dimensions [10,10,256]. Now another Maxpool 2-D layer is added with dimensions [2,2]. Output of this layer is of dimensions [5,5,256].

3. Then resultant array is flattened to be passed into Dense layer with 64 neurons.
4. All the above layers have ReLU activation function.
5. Finally, a dense layer of two neurons with SoftMax activation is used in order to get results. Threshold value of 0.5 is used to make classifications.
6. Then the model is compiled using Adam optimizer and binary cross entropy because there are only two classes.

11.3.1.4 Predictions

The model which is saved in ".h5" format is loaded and a "predict" function is made which will take new images as input and make predictions. It denotes "0" for "depressed" and "1" for "not depressed".

11.3.2 CASE STUDY 2: DEPRESSION DETECTION USING NATURAL LANGUAGE PROCESSING

A deep learning model using Recurrent Neural Networks (RNNs) is created which can take text as input and based on the given input, can classify/predict/give a small indication as to whether the person could be depressed or not depressed.

```
In [53]: model.compile(optimizer='adam',loss='binary_crossentropy',metrics=['accuracy'])
```

FIGURE 11.4 Compilation of the model.

```
from keras.models import load_model
import numpy as np
from keras.preprocessing import image

from keras.models import Model, load_model

model = load_model('image_model.h5')

def predict(photo):
    test_image = image.load_img(photo, target_size = (32,32))
    test_image = image.img_to_array(test_image)
    test_image = np.expand_dims(test_image, axis = 0)
    result = model.predict(test_image)

    predicted_class_indices=np.argmax(result,axis=1)
    labels = {'depressed': 0, 'not depressed': 1}
    labels = dict((v,k) for k,v in labels.items())
    prediction = [labels[k] for k in predicted_class_indices]
    return prediction[0]
```

FIGURE 11.5 Making predictions on image.

11.3.2.1 Dataset

Tweets' dataset from Kaggle is used for training recurrent neural network. The dataset is available on the given link: https://www.kaggle.com/kazanova/sentiment140.

Dataset used in the model consists of two columns. The first column consists of different statements, and the second column contains labels for each of these statements, depending on emotion. In this case, 0 is for "Depressed" and 1 is for "Not Depressed".

11.3.2.2 Preprocessing

The textual data obtained is not yet ready to be fed to the neural network and thus, prior to being fed to the network, preprocessing of data is required.

From the dataset all columns except the label (target) and text (tweet) are dropped as they are not useful for meaningful classifications. Also, for ease of computation, all neutral tweets are not used in the final data.

Preprocessing of data is done in the following steps:

1. First, every statement is passed through a function which removes links (URLs), images (URLs), emojis, and hashtags.
2. Then the statements are passed through a function which expands contractions (e.g., "I'll've": "I will have").
3. Then the statements are passed through a filter that removes numeric character (if any), punctuation (if any), and special characters.
4. Finally, the entire dataset is lemmatized to remove stem words present in the dataset.

```
In [1]:  import pandas as pd
         import numpy as np
         import re
         import preprocessor as p
         import matplotlib.pyplot as plt
         import ftfy

In [1]:  from keras.preprocessing.text import Tokenizer
         from keras.preprocessing.sequence import pad_sequences
         from sklearn.model_selection import train_test_split

In [14]: df = pd.read_csv("new_tweets.csv")

In [15]: df.data=df.data.astype(str)

In [16]: tokenizer = Tokenizer(num_words=70000,split=' ')

         tokenizer.fit_on_texts(df['data'].values)

         X = tokenizer.texts_to_sequences(df['data'])

         X = pad_sequences(X)

In [30]: import pickle

         # saving
         with open('tokenizer1.pickle', 'wb') as handle:
             pickle.dump(tokenizer, handle, protocol=pickle.HIGHEST_PROTOCOL)

         # loading
         #with open('tokenizer.pickle', 'rb') as handle:
         #    tokenizer = pickle.load(handle)

In [22]: x_train, x_test, y_train, y_test = train_test_split(X,Y, test_size = 0.2)
```

FIGURE 11.6 Using tokenizer to tokenize the sentences in the dataset.

```
In [22]:  x_train, x_test, y_train, y_test = train_test_split(X,Y, test_size = 0.2)

In [19]:  max_review_length = 500

          x_train = sequence.pad_sequences(x_train, maxlen=max_review_length)
          x_test = sequence.pad_sequences(x_test, maxlen=max_review_length)

In [20]:  x
Out[20]:  array([[    0,      0,      0, ...,     39,    282,    961],
                 [    0,      0,      0, ...,     32,     12,  22312],
                 [    0,      0,      0, ...,     67,     14,   1218],
                 ...,
                 [    0,      0,      0, ...,     35,     21,    913],
                 [    0,      0,      0, ...,    123,  30107,  59439],
                 [    0,      0,      0, ...,    845,     27,    158]])

In [21]:  from sklearn.preprocessing import LabelEncoder

          Le = LabelEncoder()

          Y = Le.fit_transform(df['Label'])

In [30]:  Y.max()
Out[30]:  1

In [9]:   x_train, x_test, y_train, y_test = train_test_split(X,Y, test_size = 0.2)

In [23]:  x_train.shape
Out[23]:  (838860, 45)

In [24]:  x_test.shape
Out[24]:  (209715, 45)
```

FIGURE 11.7   Sequencing the dataset and encoding the labels.

5. Now each statement in the dataset is passed through the tokenizer (breaking paragraphs into sentences and sentences into words), followed by vectorizing the text statements into sequences of integers (each integer being the index of a token in a dictionary). It also fits the text(word) into a list of numbers individually.
6. "pad sequencing" is used to sequence each statement into the same length.

Now that the data is preprocessed, it can be fed to the RNN.

11.3.2.3   Recurrent Neural Network for Natural Language Processing

1. The sequential model api of the Keras library is used to make an RNN [18].
2. First layer is the embedding layer. It requires that the input data is encoded, so that each unique number represents a single token. This data preparation step is performed during preprocessing. Random weights are initialized in this layer and thus it will learn an embedding for the entire training dataset. Input shape of input data is [45] (in length). Each data is then converted into 70,000 embeddings. (This is a very large number because in our case, we used a very large dataset and it differs with different dataset.) Output of this layer is of shape with dimensions [45,128].
3. Next layer is the LSTM layer. In the LSTM layer, one parameter is given, i.e., input shape of the tensor is 128. Output of this layer is [45,128].
4. Three dense layers were used with shapes 128, 64, and 32, respectively. Output of the first dense layer is of dimension [45,128] (45 is the length of the input text and 128 is the shape of dense layer). This is used as the input for the next dense layer of shape being 64. Output of this layer is [45,64].

```
In [25]:  from keras.models import Sequential
          from keras.layers.core import Dense, Activation, Dropout
          from keras.preprocessing import sequence
          from keras.layers import Embedding, LSTM, SpatialDropout1D

In [26]:  model = Sequential()
          model.add(Embedding(70000, 128,input_length = 45))
          model.add(LSTM(128, return_sequences=True))
          model.add(Dropout(0.2))
          model.add(Dense(128))
          model.add(Activation('relu'))
          model.add(Dropout(0.2))
          model.add(Dense(64))
          model.add(Activation('relu'))
          model.add(Dropout(0.2))
          model.add(Dense(32))
          model.add(Activation('relu'))
          model.add(Dropout(0.2))
          model.add(Dense(1))
          model.add(Activation('relu'))


          model.summary()
```

FIGURE 11.8   RNN architecture used.

Again this is the input for the final dense layer of shape 32. Output of this layer is [45,32].

5. Four dropout layers are used before each dense layer to randomly drop neurons from the previous layer output. (In our case, it is 20%.) It is done in order to avoid overfitting.
6. All the above layers have activation functions as RELU.
7. Finally a dense layer of one neuron with RELU activation is put in order to get the result (values between 0 and 1, <0.5- negative, >0.5- positive).
8. Then the model is compiled using Adam optimizer and binary cross entropy because there are only two classes.

11.3.2.4   Predictions

The model which is saved in ".h5" format is loaded and a "predict" function is made that will take new text sentences as input and make predictions. It denotes "0" for "depressed" and "1" for "not depressed."

11.4   EXPERIMENTAL RESULTS AND ANALYSIS

The Convolutional Neural Network and Recurrent Neural Network were fit on the training data of images and text, respectively.

Both the networks were successful in their compilation and execution, which trained on the entire datasets for hours and then were able to make predictions on test images and text which were given as input during testing.

Model: "sequential"

Layer (type)	Output Shape	Param #
embedding (Embedding)	(None, 45, 128)	8960000
lstm (LSTM)	(None, 45, 128)	131584
dropout (Dropout)	(None, 45, 128)	0
dense (Dense)	(None, 45, 128)	16512
activation (Activation)	(None, 45, 128)	0
dropout_1 (Dropout)	(None, 45, 128)	0
dense_1 (Dense)	(None, 45, 64)	8256
activation_1 (Activation)	(None, 45, 64)	0
dropout_2 (Dropout)	(None, 45, 64)	0
dense_2 (Dense)	(None, 45, 32)	2080
activation_2 (Activation)	(None, 45, 32)	0
dropout_3 (Dropout)	(None, 45, 32)	0
dense_3 (Dense)	(None, 45, 1)	33
activation_3 (Activation)	(None, 45, 1)	0

Total params: 9,118,465
Trainable params: 9,118,465
Non-trainable params: 0

FIGURE 11.9   Model summary.

```
In [27]: model.compile(loss='binary_crossentropy', optimizer='adam', metrics=['accuracy'])
```

FIGURE 11.10   Compiling the model.

11.5   CONCLUSIONS

It is worth mentioning here that it is very difficult even for humans to identify fake and real emotions, so expecting that from a machine is a moonshot. At this point, there are only a few tests like the Narco which can make someone speak the truth. On the other hand, this also gives **future scope** for this project to make a model which can identify the reality of an expression. Though, there are a few Machine Learning algorithms which can do so. More advanced machine learning and deep learning models can be used for getting better results. These models can also be

```
In [1]:  from keras.preprocessing.sequence import pad_sequences
         import numpy as np

In [3]:  import pickle

         # loading
         with open('./savedModels/tokenizer1.pickle', 'rb') as handle:
             tokenizer = pickle.load(handle)

In [5]:  from keras.models import load_model
         model=load_model('./savedModels/final1.h5')

In [26]: def predict(text):
             txt = []
             txt.append(text)
             text = tokenizer.texts_to_sequences(txt)
             text = pad_sequences(text, maxlen=45, dtype='int32', value=0)
             sentiment = model.predict(text)[0]
             if(np.max(sentiment) < 0.5):
                 print("negative")
             else:
                 print("positive")
```

FIGURE 11.11   Making predictions on text.

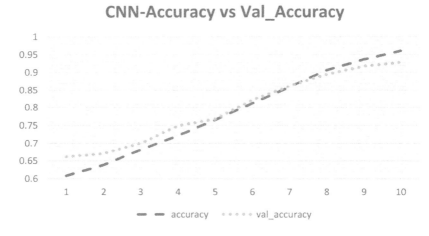

FIGURE 11.12   The CNN model gave an accuracy of 96.05% and a validation accuracy of 92.88% after 10 epochs.

modified in future to apply them on a large-scale and can prove to be one of the most significant improvements in using Artificial Intelligence in the field of healthcare.

The deep learning models can be deployed through web and mobile applications so that people can use them first hand to get an idea whether they may be depressed or not. More features can be added by coordinating and working with medical experts. For example, suggesting resources to the user like movies, books, songs, or suggesting medical treatments, according to the advice of doctors or other experts in this field. The platform (web or mobile) can be made more interactive by adding chatbots, etc. so that the application can emulate a companion and provide some kind of warmth which can motivate the user to step out from the state of continuous decline and actually seek medical help or share their problems with others or their close ones.

Therefore, in a nutshell, the models can be improved to a great extent in the future with an effort to add above-mentioned functionalities, and also train on much larger

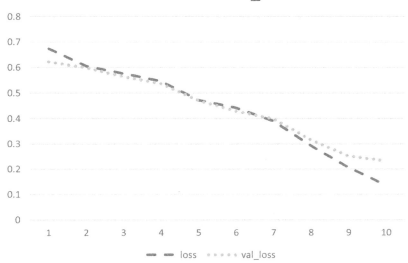

FIGURE 11.13 Obtained graph of loss and validation loss for the CNN.

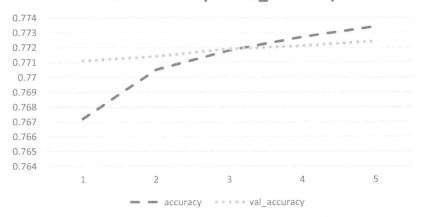

FIGURE 11.14 The RNN model gave an accuracy of 77.34% and a validation accuracy of 77.20%.

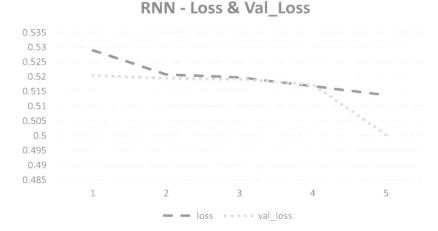

FIGURE 11.15 Obtained graph of loss and validation loss for the RNN.

datasets trying to cover different corner cases because the human emotions cannot be contained in a small list.

REFERENCES

1. Li, G., Muller, M., Thabet, A., & Ghanem, B. (2019). DeepGCNs: Can GCNs go as deep as CNNs? In *Proceedings of the IEEE/CVF International Conference on Computer Vision* (pp. 9267–9276). seoul, Korea, IEEE
2. Cho, K., Van Merriënboer, B., Gulcehre, C., Bahdanau, D., Bougares, F., Schwenk, H., & Bengio, Y. (2014). Learning phrase representations using RNN encoder-decoder for statistical machine translation. arXiv preprint arXiv:1406.1078.
3. Sindagi, V. A., & Patel, V. M. (2018). A survey of recent advances in CNN-based single image crowd counting and density estimation. *Pattern Recognition Letters*, 107, 3–16.
4. Shen, Y., Ferdman, M., & Milder, P. (2016, August). Overcoming resource underutilization in spatial CNN accelerators. In *2016 26th International Conference on Field Programmable Logic and Applications (FPL)* (pp. 1–4). IEEE.
5. Salem, M., Taheri, S., & Yuan, J. S. (2018, October). ECG arrhythmia classification using transfer learning from 2-dimensional deep CNN features. In *2018 IEEE Biomedical Circuits and Systems Conference (BioCAS)* (pp. 1–4). IEEE.
6. Chua, L. O., & Roska, T. (1993). The CNN paradigm. *IEEE Transactions on Circuits and Systems I: Fundamental Theory and Applications*, 40(3), 147–156.
7. Tang, G., Jing, M., Zeng, X., & Fan, Y. (2019, December). Adaptive CU split decision with pooling-variable CNN for VVC intra encoding. In *2019 IEEE Visual Communications and Image Processing (VCIP)* (pp. 1–4). IEEE.
8. Akhtar, N., & Ragavendran, U. (2020). Interpretation of intelligence in CNN-pooling processes: A methodological survey. *Neural Computing and Applications*, 32(3), 879–898.
9. Nakahara, H., Fujii, T., & Sato, S. (2017, September). A fully connected layer elimination for a binarizec convolutional neural network on an FPGA. In *2017 27th International Conference on Field Programmable Logic and Applications (FPL)* (pp. 1–4). IEEE.

10. Hershey, S., Chaudhuri, S., Ellis, D. P., Gemmeke, J. F., Jansen, A., Moore, R. C., ... Wilson, K. (2017, March). CNN architectures for large-scale audio classification. In *2017 IEEE International Conference on Acoustics, Speech and Signal Processing (ICASSP)* (pp. 131–135). IEEE.

11. Leevy, J. L., Khoshgoftaar, T. M., & Villanustre, F. (2020). Survey on RNN and CRF models for de-identification of medical free text. *Journal of Big Data*, 7(1), 1–22.

12. Zeyer, A., Doetsch, P., Voigtlaender, P., Schlüter, R., & Ney, H. (2017, March). A comprehensive study of deep bidirectional LSTM RNNs for acoustic modeling in speech recognition. In *2017 IEEE International Conference on Acoustics, Speech and Signal Processing (ICASSP)* (pp. 2462–2466). IEEE.

13. Shanmuganathan, V., Yesudhas, H. R., Khan, M. S., Khari, M., & Gandomi, A. H. (2020). R-CNN and wavelet feature extraction for hand gesture recognition with EMG signals. *Neural Computing and Applications*, 32(21), 16723–16736.

14. Cho, K., Van Merriënboer, B., Gulcehre, C., Bahdanau, D., Bougares, F., Schwenk, H., & Bengio, Y. (2014). Learning phrase representations using RNN encoder-decoder for statistical machine translation. arXiv preprint arXiv:1406.1078.

15. Graves, A. (2012). Long short-term memory. In *Supervised Sequence Labelling with Recurrent Neural Networks* (pp. 37–45). Springer, Berlin, Heidelberg.

16. Deshpande, M., & Rao, V. (2017, December). Depression detection using emotion artificial intelligence. In *2017 International Conference on Intelligent Sustainable Systems (ICISS)* (pp. 858–862). IEEE.

17. Albawi, S., Mohammed, T. A., & Al-Zawi, S. (2017, August). Understanding of a convolutional neural network. In *2017 International Conference on Engineering and Technology (ICET)* (pp. 1–6). IEEE.

18. Yin, W., Kann, K., Yu, M., & Schütze, H. (2017). Comparative study of CNN and RNN for natural language processing. arXiv preprint arXiv:1702.01923.

Index

Lightning Source UK Ltd.
Milton Keynes UK
UKHW050930211222
414112UK00029B/35